Advances in Oil and Gas Exploration & Production

Series Editor

Rudy Swennen, Department of Earth and Environmental Sciences, K.U. Leuven, Heverlee, Belgium

The book series Advances in Oil and Gas Exploration & Production publishes scientific monographs on a broad range of topics concerning geophysical and geological research on conventional and unconventional oil and gas systems, and approaching those topics from both an exploration and a production standpoint. The series is intended to form a diverse library of reference works by describing the current state of research on selected themes, such as certain techniques used in the petroleum geoscience business or regional aspects. All books in the series are written and edited by leading experts actively engaged in the respective field.

The Advances in Oil and Gas Exploration & Production series includes both single and multi-authored books, as well as edited volumes. The Series Editor, Dr. Rudy Swennen (KU Leuven, Belgium), is currently accepting proposals and a proposal form can be obtained from our representative at Springer, Dr. Alexis Vizcaino (Alexis.Vizcaino@springer.com).

More information about this series at http://www.springer.com/series/15228

Stanislav Ursegov · Armen Zakharian

Adaptive Approach to Petroleum Reservoir Simulation

Stanislav Ursegov ⓘ
Center for Hydrocarbon Recovery
Skolkovo Institute of Science and
Technology (Skoltech)
Moscow, Russia

Armen Zakharian
Cervart Ltd.
Moscow, Russia

ISSN 2509-372X   ISSN 2509-3738   (electronic)
Advances in Oil and Gas Exploration & Production
ISBN 978-3-030-67476-2   ISBN 978-3-030-67474-8   (eBook)
https://doi.org/10.1007/978-3-030-67474-8

This Springer imprint is published by the registered company Springer Nature Switzerland AG
The registered company address is: Gewerbestrasse 11, 6330 Cham, Switzerland

Preface

It is generally recognized that mathematical or, in other words, geological and hydrodynamic modeling is one of the most effective ways to study and forecast the development of petroleum reservoirs.

There are many different classifications of mathematical models, including their division into deterministic, stochastic, and adaptive. When using the deterministic models (from Latin—to define), mathematical equations included in them are considered fully known. Such models are characterized by the fact that accurate knowledge of their parameters in the studied interval allows us to fully determine the dynamics of petroleum reservoir development beyond the known parameters.

The stochastic models (from Latin—to guess) take into account a probabilistic nature of petroleum reservoir development, since mathematical equations used and parameters included in the models are only partially known and characterized by insufficient certainty.

The adaptive models (from Latin—to fit to), as a rule, work without any analytical equations and use various methods of machine learning in the form of artificial neural networks or fuzzy logic functions, which are first used to identify statistical patterns hidden in the collected data and then form a model mathematical apparatus to forecast petroleum reservoir development indicators on this basis. The adaptive geological and hydrodynamic models are the subject of our book.

Mathematical modeling of petroleum reservoirs has been practiced for more than fifty years, and at first glance, it may seem a purely specialized task, which is primarily engaged in reservoir and simulation engineers and which is completely uninteresting for non-specialists in this area. However, as in any narrow task, there are also universal moments that can be attractive to almost everyone. That is why in our book, we do not want to bore our readership with many important but uninteresting mathematical details, and at the same time, we will try to reveal the universal moment that is present in the geological and hydrodynamic modeling of petroleum reservoirs. This is, first of all, a contradiction between the deterministic and adaptive approaches for modeling, which can be often found in other areas of human activity that are quite far from building geological and hydrodynamic models of petroleum reservoirs.

The main goal of any simulation is to predict the future. For this purpose, three approaches are used, which differ from each other by the underlying mathematical models.

The deterministic approach, which was written about by R. Descartes, suggests that with a sufficient information base of past data it is possible to build an accurate mathematical model to forecast the future. Apologists of this direction usually attribute unsuccessful forecasts to the lack of a sufficiently complete information base, but not to the lack of the possibility of creating a deterministic forecast model.

The second approach can be defined as probabilistic, but it easily turns into deterministic if one tries to rigidly set algorithms for calculating the probabilities taken into account in such models of parameters.

In order to avoid this danger, we follow the third approach—the adaptive modeling. Its fundamental difference from the previous two is that it does not use any pre-known mathematical equations, but forms its own models based on available data and some invariant constraints. Certainly, with a sufficiently large number of such restrictions, it is possible to turn also and the adaptive approach into the deterministic, so nothing is simple here.

In our book, we want to show how to build an adaptive modeling and forecasting system using examples of petroleum reservoirs. This is a favorable area, because it has a sufficient amount of evidence and a high degree of uncertainty, which is found in most problems of interest to humanity, so the theoretical points found here can be applied to other areas. First, we would like to move the process of geological and hydrodynamic modeling, which has been standing still for a long time, and point out a new road for it, which many reservoir and simulation engineers are even afraid to look at, so as not to see the obvious disadvantages of the deterministic modeling techniques of petroleum reservoirs.

The founder of the theory of fuzzy sets, L. Zadeh, formulated a known principle of incompatibility, the essence of which is that the more complex the system, the less we are able to give accurate and at the same time practical judgments about its behavior. For systems whose complexity exceeds a certain threshold, accuracy and practical meaning become almost mutually exclusive characteristics. Thus, the consequence of the principle of incompatibility can be expressed as follows: Accurate quantitative methods are not suitable for modeling such complex systems as petroleum reservoirs in the process of their development, so the non-traditional approaches must be used as forecasting tools, for example, the adaptive, based on the analysis of collected data and machine learning algorithms.

Our book consists of an introduction, seven chapters, and a conclusion. The introduction shows that the way out of the impasse in which deterministic geological and hydrodynamic modeling of petroleum reservoirs is associated with adaptive modeling, which has real prospects for further development.

The question of the initial information capacity for geological and hydrodynamic modeling is considered in Chap. 2. Here are the answers to the questions of how much initial information actually gets into the models, based on the classical theorem of C. Shannon, what uncertainties arise in this case and how they determine the feasibility of complicating the models.

Chapter 3 explains what an adaptive model is and how it differs fundamentally from a deterministic one. The principle of adaptability was first formulated by W. Ashby, and this chapter shows how it can be applied in geological and hydrodynamic models and why this is not the case in deterministic models.

Chapter 4 is devoted to choosing machine learning algorithms for adaptive models. Recently, much has been said about artificial neural networks as a universal tool for analysis and prediction. In this chapter, we will talk about our own experience with artificial neural networks, gained over more than ten years, and why they had to be abandoned in favor of fuzzy logic algorithms, as well as will suggest new forms of artificial intelligence based on fuzzy logic and how they are applied in the practice of adaptive modeling.

Chapter 5 discusses the difference between an adaptive geological model and a deterministic one, which is based on the manual work of geologists in the pre-computer era. Today, this has definitely outlived its usefulness. The computer implements other methods. Therefore, it is necessary to break away from the deterministic tradition in order to move on. The main purpose of an adaptive geological model is not to display actual information about the structure of a petroleum reservoir as accurately as possible, but to predict this structure in areas that are not yet drilled. The adaptive approach proves that in order for a geological forecast to be successful, it is not necessary to obsequisoly adhere to the facts, since this is often harmful and the geological structure of the reservoir at the points where the wells are located is known without modeling. Just as in regression analysis, it is not necessary for the function to pass exactly through all the actual points, so it is not necessary for the geological model of the reservoir to reproduce the actual data completely. This is a fundamental feature of the adaptive geological model.

Chapter 6 is devoted to the questions of hydrodynamic modeling of petroleum reservoirs. It is believed that deterministic hydrodynamic models are based on partial differential equations, although in fact this is not quite true, since the correctness of the transition to finite-difference equations cannot be justified. On the other hand, the actual geological structure and fluid filtration process in petroleum reservoirs cannot be described by partial differential equations. The apparatus of percolation theory and cellular automata is more suitable for this purpose. The essence of the adaptive hydrodynamic model is that it is based on just such an apparatus and this is its main advantages. Using an adaptive hydrodynamic model, one can get the same result as using a deterministic hydrodynamic model, but at a much lower cost.

Chapter 7 deals with universal problems of forecasting. An adaptive hydrodynamic model is not needed by itself, but for predicting the results of the development of the petroleum reservoir for which it was created. This chapter presents a method of adaptive forecasting of development indicators, effects of well workover measures, and oil and gas production levels, which has been used for almost twenty years. Some people do not understand that forecasting is an inverse problem that does not have a single solution. This is why, it is impossible to successfully predict any single result, for example, an oil rate of a single well. Nevertheless, this does not mean that forecasting does not make any sense. The chapter shows objective limitations of forecasting and how such forecasting can be useful.

Chapter 8 should not be considered as a conventional advertisement of the Cervart software system that we have created. The main purpose of this chapter is to demonstrate that the proposed version of adaptive modeling is not a purely theoretical invention, but it is numerically implemented and has more than a decade of practical experience in various petroleum reservoirs.

In conclusion, our book emphasizes that no one problem can be solved for the last time. They are usually infinite. Therefore, we list the problems that we have not been able to solve now and explain why they are as important as the issues that have already been resolved. It is important to understand what to look for at the next stage, and only then, it can be found in principle.

In different years, Profs. B. Sokolov, Yu. Kosygin, Az. Mirzajanzade, V. Lysenko, I. Gutman, S. Zakirov, L. Ruzin, Y. Konoplev, E. Gildin, and I. Y. Akkutlu took part in conducting and discussing the results of adaptive modeling presented in the book. We express our sincere gratitude to all of them and many other specialists, whose fruitful work allowed us to write this book.

Naturally, our book has flaws that need to be discussed. At the same time, it is desirable to take advantage of the wise advice of J. d'Alembert: If criticism is fair and benevolent, it deserves gratitude and respect; if it is fair but devoid of benevolence, it only deserves respect without gratitude; if criticism is un fair and unfriendly, we will pass it by in silence and put it into oblivion".

Moscow, Russia
August 2020

Stanislav Ursegov
Armen Zakharian

Contents

1 Introduction 1
    References 2

2 Information Capacity of Initial Data 3
    References 17

3 Contrasts Between Adaptive and Deterministic Models 19
    References 25

4 Alternatives for Mathematical Apparatus of Adaptive
    Simulation—Neural Networks and Fuzzy Logic Functions . . . 27
    Reference 36

5 Adaptive Geological Modeling 37
    References 49

6 Adaptive Hydrodynamic Modeling 51
    References 60

7 Adaptive Forecasting 61
    References 73

8 Adaptive Software System Cervart 75

9 Conclusion 83
    Reference 84

Index 85

If you shut your door to all errors,

Then the truth will be shut out.

Rabindranath Tagore
(1861–1941)

The truth is so good. It gives everyone the necessary point of support for self-esteem in the shaky attitudes of life. It is so tempting to reject everything that does not coincide with the idea of the truth of the democratic majority. However, Hegel expressed that neither universal approval, nor collected facts, were the absolute criteria of the truth; the only criterion of the truth was the logical consistency of its proof [1]. The meaning of this statement in relation to this book is that even if everyone around us will say that the deterministic geological and hydrodynamic modeling of petroleum reservoirs is the only correct approach—this cannot be the proof of this statement's truth.

Moreover, if we talk about facts, these are the contradictory things in the world, although they are treated as something incredibly reliable. However, everyone knows how during trials, the prosecutor and the lawyer easily juggle the same facts to prove each other right. Because every fact in itself is not something simple and elementary. It is described by large amount of information, sometimes very contradictory, which makes it possible to maneuver.

What truth can be in the geological and hydrodynamic modeling of petroleum reservoirs? After all, this is just a game, because any model is just a toy, no matter how seriously, it is taken. Certainly, we understand the game within the framework of the game theory, formulated by Von Neumann and Morgenstern [2]. This theory was necessary because it was impossible to describe economic behavior using the deterministic methods since the behavior was too complex, the initial information was not enough, and it turned out to be confusing and ambiguous. The same can be said about the development process of petroleum reservoirs. It cannot be described only by the deterministic approach.

In [2], Von Neumann and Morgenstern also formulated the theory of the expected utility, which allowed estimating the profit correlated with the costs. Let us assume that neither deterministic nor adaptive model makes the profit, since the Securities and Exchange Commission does not accept the results of its calculation as reserves of the reservoir under study. Definitely, there will be "scientific heads" ready to do anything to prove that there is some "reliable"

© The Author(s), under exclusive license to Springer Nature Switzerland AG 2021
S. Ursegov and A. Zakharian, *Adaptive Approach to Petroleum Reservoir Simulation*,
Advances in Oil and Gas Exploration & Production,
https://doi.org/10.1007/978-3-030-67474-8_1

simulation, the results of which can be accepted, but this is not true.

In the existing deterministic simulators, there are many levers, which can be used to get any predetermined result. The deterministic simulator cannot be objective by its nature, because it is designed for a specialist. While the adaptive simulator proposed in this book is a form of artificial intelligence to which the specialist does not have any access. If the profit from both simulators is zero, then the expected utility is determined by the costs, and they are an order of magnitude less for the adaptive simulator than for the deterministic.

Why is there a need for this book about the adaptive simulation of petroleum reservoirs? First, to tell the readership that the methods of deterministic modeling that are currently widely used are neither the only possible nor the only correct, and that the readership should pay attention to other approaches.

The deterministic geological and hydrodynamic modeling has almost half a century of history. Tens of thousands of specialists around the world are engaged in it. Many scientific conferences are organized annually. A lot of papers and books are published that address the issues of modeling methodology. All of this ultimately resembles a kind of cult, deviation from which is seen as heresy, because there is no other way and there cannot be. However, all this has come to a standstill, and in recent years, it has not had any significant development. Geological and hydrodynamic models are becoming more and more cumbersome, containing up to a billion cells. However, this does not seem to be of much use, since the bulkiness of models is not traditionally compared with the quantity and quality of initial information available for their construction.

In this regard, the book shows that there is a way out of this situation and it opens up prospects for future development. Perhaps this is the proposed adaptive model, perhaps not, and there are other solutions, but the main thing is not to turn the geological and hydrodynamic modeling into an absolute truth.

This book does not present many formulas. It has only the most basic ones. First, because the basic method of adaptive modeling is machine learning, which is difficult to represent using analytical expressions. Second, most hydrodynamic modeling formulas simply migrate from one book to another just to give the new publication a scientific appearance. The entire mathematical apparatus of reservoir simulation is described in detail in the now classical book of Aziz and Settari [3].

It is unlikely that any of the specialists working on simulators know all the details of this mathematics, but even fewer of those who besides developers of deterministic modeling know how such simulators actually operate. For such complex systems, there are always tricks to make ends meet, and they are not usually described in books. Therefore, it is important to present the main idea, and not hide behind formulas, unless, of course, the book is not on mathematics and its topic is not the conclusion of these formulas themselves.

References

1. Hegel, G. W. F. (1969) *Science of logic* (A. V. Miller, Trans.). London: George Allen & Unwin Ltd; New York: Humanities Press.
2. Von Neumann, J., & Morgenstern O. (1944) *Theory of games and economic behavior*. Princeton University Press.
3. Aziz, K., & Settari, F. (1979). *Petroleum reservoir simulation*. Netherlands: Springer.

It is obvious that the development process of any petroleum reservoir is subject to physical laws. Followers of the deterministic modeling approach prefer to emphasize that their method is based on the implementation of the partial differential equations, and it gives their method a superiority to the others, for example, the adaptive. However, for any type of modeling, it is important not only using physical laws, but also understanding that the development process is known only as much as there is information about it.

What is the meaning of a physical law expressed using a mathematical formula—that it creates a variety of values of a target parameter (for example, a liquid rate). This is its informational capacity. If nothing is known about the target parameter, it is a fair assumption that all its values are the same. The available information allows differentiating the values of the target parameter, i.e. to create a variety of them. Such information can be obtained using a mathematical formula or by observing changes in the target parameter.

There are target parameters that are important from a physical point of view, such as a reservoir pressure, but from the informational point of view, they may not be of great importance, because these parameters, like the reservoir pressure, are rarely measured and with large errors. Therefore, when modeling, it makes little sense to pay attention to such parameters based only on physical principles. The reservoir pressure is certainly an important parameter, but there is not enough information about it.

If one looks at the Dupuit's equation: $Q = \frac{kh}{\mu}\Delta P \big/ \ln\left(\frac{R_d}{r_w}\right)$, where Q—liquid rate, k—absolute permeability, h—net pay thickness, ΔP—difference between reservoir and wellbore pressures, R_d—drainage radius, r_w—wellbore radius, all its parameters are represented by real numbers. From it, one can get a large variety of values of the liquid rate, much more than the real amount of information that can be obtained from observing the actual liquid rates. The main problem is that all the parameters included in the formula do not have any informational capacity and cannot be represented by real numbers, but only in an interval scale, i.e. it is not correct to apply this formula.

In the literature concerning the information theory, everything usually comes from the Shannon's entropy [1]. However, there is an interesting statement "information is physical" [2], which is hardly supported by anyone due to insufficiently strict proof. "And then, the question arises: is the reductionist slogan "information is physical" a latest definition of physics, one that chooses to further expand physics by incorporating phenomena related to information, or on the contrary, that slogan is a mere claim in which the concepts of "physical" and "information" are only assumed to be denied in some vague and tacitly accepted ways?" [3]. But every path begins with the first step and here is the next assumption that seems also useful: "Indeed, as it turns out, information is in fact so fundamental that the whole of quantum mechanics can be reconstructed from no more than three axioms with clear empirical motivation, the first of which is called information capacity: All systems with information carrying capacity of one bit are equivalent" [4].

The question about the information capacity of geological and hydrodynamic models is not considered too often because of the fear that if it is strictly considered, it can be concluded that there is no real sense in huge petroleum reservoir models, since there is not enough information to build them. A number of papers, for example, consider the question of the amount of information in geological and hydrodynamic models as such [5–7]. This is done using the information entropy formula of Shannon: $H(X) = -\sum_{i=1}^{n} P(x_i) \log P(x_i)$, where H—entropy, X—variable, P—probability, x—outcome, by constructing dozens or even hundreds of model realizations with different values of their parameters in the grid cells. This approach cannot be considered correct, because each realization occurs by the implementation of different construction methods, but not changes of the initial information. The variation resulting from the variety of construction methods is outside the frame of this book. In any case, such variation only reduces the reliability of the model since different realizations can be obtained from the same initial data set.

The purpose of this chapter is to show how much of the initial information actually appears in the reservoir models, based on the seventh theorem of Shannon, what uncertainties exist here and how they determine the feasibility of complicating the models.

As in no other area, the reservoir simulation has long been and widely used in the petroleum industry because any reservoir has an obvious geometric shape that allows specifying its modeling task. Previously, reservoir models were 2D and created manually on paper. Now, a computer builds 3D reservoir models, and they contain hundreds of millions of cells, but the quality of

the reservoir simulation has not changed. The 3D reservoir model primarily displays the reservoir's geometric shape and the heterogeneity of its rocks. However, the amount of initial information for constructing such models has stayed the same. As before, there are only seismic data that can be used to create the structural surface of the reservoir, and data on drilled wells that give an idea of the lithological and petrophysical heterogeneities of the reservoir rocks. Obviously, the 3D seismic survey has more information capacity and better than the previous 2D one, because it has a much larger and uniform network of profiles. The quality of well logging has also increased. However, wells form only a small subset of points in the total set of reservoir points for which there are no data. Such situation has not changed since drawing the 2D reservoir maps.

The 2D model itself contains less information than the 3D computer model. However, the amount of input information is the same, so for every bit of information of the 2D model, there is significantly more input information and, therefore, the 2D model is more reliable than the 3D. This is confirmed by the practice of reservoir simulation, when first, the 2D model is built, and then the 3D model is modified according to the previous. In other words, the available amount of initial information is only enough to build the 2D model. Otherwise, it would be the opposite, and the 2D model would be obtained as a generalization of the 3D model. It is possible to suggest a more general statement—the 3D model contains no more information than the 2D, since both of them have the same amount of initial information.

According to the seventh theorem of Shannon: "The output of a finite state transducer driven by a finite state statistical source is a finite state statistical source, with entropy (per unit time) less than or equal to that of the input. If the transducer is non-singular they are equal" [1]. The reservoir model can be considered as a singular transducer of initial information, since it is based on seismic and well data through various interpolation methods, but it is not possible to get these data back from the model.

Consider the following schematic example. There is a 1D grid with a size of 10 by 10, i.e. only 100 cells, and there are 10 wells whose data should be interpolated to this grid (Fig. 2.1). Let us assume that it is about the gross thickness of a petroleum reservoir, varying from 1 to 10 m and which is known with an accuracy of 1 m. This gives the amount of information equal to 3.5 bits per well (the logarithm of 10 on base 2). In addition, there is information about the coordinates of wells, which are determined with accuracy to a cell number, also from 1 to 10, which gives 3.5 bits for the X coordinate and 3.5 bits for the Y coordinate. In total, there are 10.5 bits per well and 105 bits per 10 wells. This information is distributed over 100 cells—with an average of 1.05 bits per cell. At the same time, there is no information about coordinates of the cells, since they are located on a regular grid, so all information relates only to the value of the gross thickness of the reservoir in a cell, which on average can be known with an accuracy of 5 m, i.e. it is greater than 5 m or less.

If the grid spacing is refined, the number of cells will increase up to 400, but the initial information will only increase due to the information about the coordinates of wells from 7 to 14 bits per well—in total there will be 175 bits of input information or 0.43 bits per cell. The amount of initial information per cell has decreased, because the parameter value in neighboring cells does not vary. Continuing to refine the grid, one can get that in the limit, one cell will contain 0 bit of information.

If we understand the model as mathematical, then the first question that can be asked for any cell is whether it belongs to a subset of cells of the reservoir under study itself or not. In principle, the cell can be located above the top or below the bottom of the reservoir; although in the model it is understood as belonging to the reservoir.

All model cells can be divided into the following two subsets: the first one is relatively small; these are the cells through which the wells pass through, the second—all the others. And here comes the first uncertainty—it is impossible to accurately determine the cells through which a particular well passes. The model cells are usually 25–50 m in size, and the most wells are deviated and extend 200–400 m from their heads. If we consider the errors of wellbore survey within the tolerance of 2° in azimuth, then at a depth of 1500 m, the uncertainty of the position of the well into the reservoir at the coordinates X and Y is about 12 m. If the grid has a size of 25 m, then there is a probability that the well will not be in the cell through which it passes according to the calculation, but in one of the neighboring ones. To do this, it is enough to calculate the well point to be at a distance of just over 2 m from the center of the cell. This gives a 97% probability that the well will fall into one of the neighboring cells. With a 50 m grid, this probability is much less and is only 65%, and with a 100 m grid, it is only 23%. The limit at which this probability becomes less than 50% is 71 m. Thus, the accuracy of the wellbore survey limits the opportunity for refining the grid spacing.

If a well passes through a cell, then this cell can be assigned the value of some geological parameter recorded in the well, for example, the absolute depth of the top or the gross thickness of the reservoir. While in neighboring cells, the parameter value is already obtained by

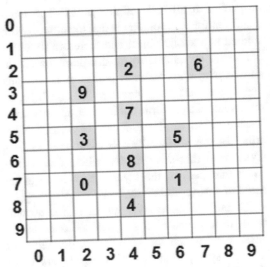

Fig. 2.1 Interpolation scheme of 1D model with 100 cells and 10 wells

interpolation and may differ from the value recorded in the well assigned to the cell through which this well passes. However, if there is uncertainty in the position of the well on the model grid, then there is no point in trying to ensure that the cell through which the well passes is assigned a geological parameter of this well.

The wellbore survey error also affects the absolute depth of the top of the reservoir under study. With a tolerance of 0.5°, the error at a depth of 1500 m can reach 2.5 m in an absolute value. Nevertheless, more significant is the uncertainty of the position of the vertical point, which is taken as the top of the reservoir. As a rule, the reference point for which it is allocated is not maintained over the entire area of the reservoir. At best, the uncertainty of the top position is about 2 m. This is difficult to justify, but it is known that two versions of stratigraphic markers made by different geologists almost never completely coincide.

It is possible to assess the uncertainty of the reservoir top position by analyzing the coherence of permeable sublayers identified by logging in neighboring wells. By estimating the position of these sublayers by the distance from the selected top or bottom of the reservoir, one can calculate the coefficient of coherence based on the proportion of coincidence of permeable and impermeable sublayers in these wells. It is assumed that this coefficient should be the maximum for correctly selected layer boundaries. By consistently shifting the boundaries in the range of 2–3 m, one can find the position at which the coherence coefficient is maximal. Often the coefficient is approximately the same in a certain range of displacements, and this interval just shows the uncertainty of the position of the layer boundaries. When the gross thickness of the layer in neighboring wells is close to each other, this uncertainty is about 2 m. If these thicknesses differ significantly, the uncertainty increases in proportion to the difference in thickness.

This uncertainty shows that the "slicing" of thin layers by 0.4 m in the geological model is most often incorrect due to the uncertainty of the position of the layer boundaries. When the layers are "sliced", the interpolation is mainly based on them according to the commonly used anisotropy coefficient. It is more correct to take the thickness of layers equal to the double uncertainty interval—about 4 m. However, there are also problems when the gross thickness of the layer changes significantly within the limits, which is usually observed. In this case, it is not always clear what to do—change the thickness of the layers in proportion to the gross thickness of the reservoir or interrupt the upper layers, assuming the erosion of the reservoir. One can also use a mixed strategy—somewhere to reduce the thickness of layers, and somewhere to interrupt them. However, this already resembles the process of detailed correlation and requires the individual participation of a geologist, whose qualification does not always correspond to this task. Usually, one specialists perform detailed correlation, and others, who are mainly engaged in obtaining the 3D model that does not contradict the previously constructed 2D model, perform the construction of the 3D computer model.

Preliminary construction of the 2D model actually makes sense because this model fully takes into account the uncertainty of geological information about the reservoir. The main thing is that the boundaries of the layer are manually selected and therefore the layer is a single subsystem with its own development patterns, and therefore interpolation can be used to restore its properties in all cells of the model. The main condition for interpolation is that the object being studied is a single object and therefore its properties can be calculated at any point based on known properties at other points. It is obvious that if one uses the parameters of one layer to interpolate the properties of another, for example, located below or above, it will be wrong. Nevertheless, this is often the case when interpolating layers in a thin-layer model, where due to the small thickness of the layers, there is no guarantee of their correlativity.

To build a multi-layer model correctly, one needs to have the results of detailed correlation, because of which the reservoir is divided into several layers. That is how many layers there should be in the 3D geological model. Then this will be correct, since it is certain that the interpolation is performed within the same layers. It is desirable that these layers are selected manually. Because in this case, the model receives new information about the geological structure of the reservoir. Fully automatic detailed correlation is very questionable. Detailed correlation is probably the only thing that cannot be done with a computer in the process of geological modeling.

Let us consider the specific amounts of information contained in the geological and field data on the example of a very large petroleum reservoir in the Russian Western Siberia oil and gas bearing province, where several thousand wells have been drilled and therefore there is a fairly stable statistical basis. Specific examples of actual information are more obvious than general conclusions drawn from hidden data.

If one needs to know the value of any geological and field parameter of the well—stratigraphic markers, petrophysical parameters, liquid rate, water cut or pressure, then one needs to get some information about it. In the classic definition of Ashby [8], information is understood as a measure of diversity. The more values a parameter can take, the more information is needed to know its specific value in each case. It should be noted that the value of information entropy according to the Shannon's formula depends on how many intervals the sample is divided into. If it is divided, for example, into 16 intervals, the maximum entropy will be 4 bits per sample line, and if it is divided into 32 intervals, then, respectively, 5 bits. In principle, the number of split intervals should be determined by the measurement accuracy, which is not always possible to determine. One can take the accuracy of how data is recorded in the database, although this is usually incorrect and the actual accuracy is significantly lower than the recorded one. There is no dispute that the geological and field parameters can only be measured in the interval scale of measurement, and thus all measurements that are represented in the database as fractional numbers can be reduced to natural numbers. The amount of such numbers shows how many values a parameter can take.

The amount of information required to know any parameter value according to the Hartley's formula is generally equal to: $I = \log_2 N$, where I —information capacity of the parameter, N— number of values that the parameter can accept. In this case, this is the maximum amount of information that is obtained when all parameter values are evenly distributed. For an uneven distribution, the amount of information can be calculated using the Shannon's entropy. The Hartley's formula is a special case of entropy for a uniform distribution: $I = -\sum_{i=1}^{n} P(x_i) \log P(x_i)$.

Let us start with the geological parameters of formation A. Data on the absolute depths of the top of this formation are available at 15,800 points and at a given accuracy of 0.2 m, 775 values are obtained, which gives 9.6 bits of information, and taking into account the uneven distribution—9.24 bits or 96.2% of the maximum. The total amount of information about the position of the formation top is 146,000 bit. This is much less, than the information contained in the structural map based on seismic data. At a given accuracy of 0.5 m, there are 350 values for 3,288,000 cells, which gives 8.45 bits of information, and taking into account the uneven distribution—7.82 bits or 92.5% of the maximum. As total, 25.712 Mbits of information are obtained. In order to build a structural map with an accuracy of 0.5 m, one needs to have the same amount of information, but according to stratigraphic data, there are only 146,000 bits, which is 0.56% of the required amount of information. Certainly, this is not enough and it is necessary to use seismic data along with well data to build a structural map.

The data on the gross thickness of the formation contain less information than in the absolute depths of the top. From a sample of 15,700 rows, 105 values are obtained with the same accuracy of 0.2 m, which gives 6.71 bits of information, and taking into account uneven distribution—5.65 bits or 84.2% of the maximum. In this case, information on the gross thickness is less than the absolute depths of the top of the formation because the scale of values is substantially less than the magnitude of the values of absolute depths of the formation top. One can say that it does not matter much, because that is the nature of the data. However, this will affect later, when it comes to forecasting. A less variable parameter will also give a less variable or less informative forecast.

The net pay thickness is more informative. With a sample of 12,826 rows, 156 values are obtained with the specified accuracy of 0.2 m. This gives the number of 7.28 bits of information per measurement, and taking into account the uneven distribution—6.78 bits or 93.1% of the maximum. The net pay thickness is more variable, because there may be different net pay thickness for the same gross thickness.

The structure of permeable sublayers in the formation is also interesting which can be expressed in particular in terms of the thickness of individual permeable sublayers. With a sample of 160,936 rows, 155 values are obtained for a given accuracy of 0.1 m and 7.27 bits of information per measurement, as for the net pay thickness of the formation as a whole, but taking into account the uneven distribution, only 4.34 bits or 60% of the maximum are obtained.

The porosity parameter for a sample of 160,936 rows and a specified accuracy of 0.001 gives 160 values and 7.32 bits of information, and taking into account the uneven distribution—5.8 bits or 79.2% of the maximum. However, it should be taken into account that the porosity is not directly measured, but is recalculated from the logging parameters, in this case, the PS logging values. This distinguishes porosity from measurements of net pay thickness and identifications of permeable sublayers that are the direct problems. Trust to the recalculated parameters should be the same as, say, in a criminal trial, trust to direct and indirect evidence. The latter are valued much less. Therefore, the information value of petrophysical parameters, no matter how important they are from a physical point of view, is much lower than the measurement of net pay thickness. Here there is a contradiction of the deterministic approach to hydrodynamic modeling. For this approach, the petrophysical parameters and the net to gross ratio are the same in

terms of confidence. However, this is not the case at all.

The permeability for a sample of 160,934 rows and a given accuracy of 1 mDarcy gives 240 values or 7.9 bits of information, but taking into account the uneven distribution, it turns out only 4.45 bits or 56.3% of the maximum. It is known that the permeability is recalculated from the porosity through an empirical function, so it cannot contain more information than the porosity. Nevertheless, more information is obtained because the total sample is divided into groups based on the lithotypes, which are usually selected manually and for each lithotype select their own empirical function. This is hardly correct. If it was not talking about lithotypes, but say about the content of quartz or clay cement in sandstones or the degree of grain sorting, this would be quantitative information that would contribute to the calculation of permeability. However, when manually selecting lithotypes, an abuse leads to an unjustified increase for information about the permeability and further to errors in the calculation of the hydrodynamic model. It should be understood that if the porosity by which the permeability is calculated is itself an indirect evidence or a recalculation of parameters, then the permeability is an indirect evidence squared and it can no more be trusted than the porosity. During the adjustment of a deterministic hydrodynamic model, it is understood that the permeability parameter is the least reliable, so it is usually modified.

The oil saturation parameter is more informative than porosity. With a sample of 160,589 rows and a specified accuracy of 0.001, 590 values or 9.2 bits of information per measurement are obtained, and taking into account uneven distribution—8.56 bits or 93% of the maximum. Definitely, if one reduces the accuracy to 0.002, the amount of information will decrease, but in this case, it is important that the distribution of oil saturation is closer to uniform than that of porosity and the same values are less common here.

The geological parameters form a single block of information that is used for predicting the field parameters, such as the initial oil rates of new wells. Therefore, if there are many identical values in a geological parameter, such as in permeability, and they account for different oil rates, this shows that the parameter does not have the information capacity for solving its main task —forecasting the oil rates.

The second information block is the field parameters. Let us look at the situation with the amount of information here. The main types of information in the field block are measurements of liquid rates and water cut, measurements of reservoir pressures and liquid levels, as well as data from monthly reports on oil production, which should be formed from measurements, but in fact, this is not quite true.

Let us assume that the liquid rates of formation A can be measured with an accuracy of 0.5 t/d. Certainly, this is an exaggerated accuracy, but in this case it is not the main thing. Since it is not the absolute values of the amount of information that are significant, but relative values, in the sense of which type of data contains relatively more or less information. In total, the database contains 179,292 lines of measurement of liquid rates. With a specified accuracy of 0.5 t/d, the liquid rates can take 872 possible values. Given equal probability for all values, it turns out that a single measurement can contain 9.77 bits of information, but since the values are not actually equally probable, then according to the Shannon's entropy formula, one gets only 6.81 bits or 70.0% of the maximum number.

If we take the specified accuracy of the liquid rate measurement equal to 1 t/d, we get 629 intervals of values and, accordingly, the maximum amount of information per measurement contains 9.3 bits, and for the uneven distribution —6.75 bits or 72.5% of the maximum. In general, this compares fairly well with the split at an accuracy of 0.5 t/d, so without losing generality, one cannot pay much importance to the specified measurement accuracy. It is important that the amount of information in a single measurement of the liquid rate is higher than in measurements of porosity (5.8 bits), and even more permeability (4.45 bits), but comparable to net pay thickness (6.78 bits).

Now let us estimate the amount of information in the water cut measurements. With a sample of 179,437 rows and a specified accuracy of 0.001 (for example, because the water cut values are written as a percentage with tenths), 927 values are obtained and, accordingly, the maximum amount of information per value is 9.86 bits, and taking into account the uneven distribution— 7.29 bits per measurement, or 73.9% of the maximum. If we take the measurement accuracy of 0.01, which is closer to reality, we get 100 values, which is more natural. Since water cut varies from 0 to 1 and there simply cannot be more values, 6.64 bits of information per measurement or 6.34 bits with the uneven distribution, i.e. 95.5% of the maximum. The fact that the distribution is close to uniform shows that the parameter is highly informative. Note that when one specifies an accuracy of 0.001, water cut becomes more informative than measurements of liquid rates, and it is correct, because water cut is measured in the lab, and while liquid rates are measured using a mechanical device connected to several wells of the same pad and are generally little reliable. Although on the other hand, liquid samples for measuring water cut are taken relatively rarely, several times a month, and even within one day one can get samples with significantly different water cut, so for a real assessment of water cut, the number of samples that are usually available is not enough.

As an example, data on measurements of the amount of methane in well gas samples are presented. These samples are always taken two at a time from the same outlet with a time interval of one to two minutes. However, the methane content in them is always different. 1500 samples were analyzed, and the average difference in methane content in the paired samples was 0.8% with an average methane content of 93%. In some cases, there were more than two samples, and in one case, eleven samples. For such cases, it was possible to calculate the average methane content for two, three, or more samples and estimate variations in this average. At the same time, from eleven samples, it was possible to take different samples of three, four or more measurements. It turned out that the more samples, the less variation and with eight samples, it decreased to 0.2%, but it did not become less. In order to reduce the error from 0.8 to 0.2%, one needs at least eight samples. Nevertheless, it is too expensive.

When measuring reservoir pressure, two pressure gauges are used simultaneously and there is an error associated with instrumental errors between their measurements. The average difference was about 1.5 Atm with an average pressure of 350 Atm. Variations of measurements would increase if measurements were taken several days in a row at different times of the day. Therefore, it would take six measurements to reduce the error to 0.5 Atm. Nevertheless, it is unlikely that anyone will agree to do this. This is also the case for water cut measurements. They are clearly not enough. In addition, this should be taken into account and not treat the measurement data as something actual, which must be reproduced during modeling.

If now, using measurements of liquid rates and water cut, calculate oil rates, then with the specified accuracy of 0.5 t/d, one will get 330 values with a sample of 179,292 lines and 8.37 bits of information per measurement, and taking into account the uneven distribution—6.17 bits or 73.7% of the maximum. The most important thing here is that the amount of information in the calculated oil rate is less than the information in the liquid rate measurements (9.77/6.81 bits) and water cut (6.64/6.34 bits). This conclusion fully corresponds to the seventh theorem of Shannon. Since the input had 6.81 bits of information from measuring the liquid rate and 6.34 bits of information from measuring the water cut were added to it, the result was only 6.17 bits. Now let us analyze the amount of information in the oil rate obtained from the monthly production report by dividing oil production by the working time. With an accuracy of 0.5 t/d, 984 values are obtained from a sample of 282,340 rows, which gives the amount of information 9.94 bits, and taking into account the uneven distribution—5.7 bits or 57% of the maximum. That is, the oil rate in the monthly production report contains less

information than the oil rate calculated by real measurements, as it should be.

Calculating the monthly production report data, we get 1856 values from the same sample of 282,340 lines and, accordingly, 10.86 bits of information per line, and taking into account the uneven distribution—7.72 bits or 71% of the maximum. Here, the amount of information of the liquid rate obtained from the monthly production report were more than the amount of information of the liquid rate obtained from the real measurement (6.81 bits). This is not correct, since the liquid rates in the monthly production report are not calculated based on the measurements. The larger sample size may affect this, but in principle, it confirms the existing distrust of data from monthly production reports, which are often smoothed out. Nevertheless, it is the data that the hydrodynamic models are usually history matched.

Paradoxical properties of the monthly production reports are also found if we evaluate the information capacity of the injection rates. There are 2269 values of injection rates from a sample of 61943 rows at an accuracy level of 1 t/d. This gives 11.15 bits of information per line, and taking into account the uneven distribution—8.9 bits or 87.7% of the maximum. It turns out that the data on the injection rate is more informative than the data on the oil and liquid rates. However, this does not correspond to reality, because it is well known how carelessly data is generated on injection. The apparent high information capacity rather shows how unreliable these data are.

Let us consider the information capacity of reservoir and bottom-hole pressures. For formation A, there are 26,484 measurements of reservoir pressures, which at an accuracy of 0.5 Atm give 126 values or 7 bits of information per measurement, and taking into account the uneven distribution—5.24 bits or 73.4% of the maximum. This information capacity is noticeably lower than the information capacity of liquid rate and water cut measurements, which fully reflects the real inaccuracy of reservoir pressure measurements. Here we can see that from the point of view of physical processes, reservoir pressure is a very important parameter, but, unfortunately, it does not have a very big information capacity.

A little more information is provided by measurements of bottom-hole pressures. With a sample of 51,609 rows, 165 values are obtained with an accuracy of 0.5 Atm, which gives 7.37 bits of information per measurement. At the same time, the distribution of bottom-hole pressures is close to uniform, so the amount of information per measurement is 7 bits or 95% of the maximum.

The measurements of liquid levels in wells have a higher information capacity. From one side, their sample is significantly larger—102,245 rows and it gives 1630 values with an accuracy of 1 m which results in 10.67 bits of information per measurement, and taking into account the uneven distribution—10.08 bits or 94.5% of the maximum. In other words, measurements of liquid levels in wells are more informative than reservoir pressure measurements and can be used to at least partially fill in the lack of information on reservoir pressure measurements.

Ultimately, the amount of information depends on the distribution of the analyzed parameter and largely on the main parameters of any distribution—the average and variance. Previously, we considered the amount of information in the entire reservoir, but we can first calculate the average for each well, and then the distribution and information capacity of these averages. Then, for any particular well of formation A, the variations of liquid rates will be less than for the entire formation by about four times. Note that we previously considered the amount of information depending on the parameter values for a given accuracy, but now we will use a different way and divide any sample into 128 intervals, so that the number of values is the same everywhere. Then the amount of information will depend only on the form of distribution and, above all, on the variance. Therefore, the numbers of information capacity below will differ slightly from those given earlier.

Thus, for a single well, one needs 7.2 bits of information to know the liquid rate with an accuracy of 1 t/d, or an average of 2.58 bits to know with an accuracy of 25 t/d. However, for different wells, this value is not the same: for low-rate wells, it is enough to have 2.5 bits to know the liquid rate with an accuracy of 1 t/d, and for high-rate wells—4.6 bits to know the liquid rate with an accuracy of 25 t/d. Note if a particular well requires less information, but it requires information about what kind of well it is, that with a total number of wells equal to 2500 is 11.3 bits. Although this information seems to be obtained easily, it allows reducing the amount of information that is necessary to know the liquid rate.

A specialist who knows wells by their numbers can immediately tell the order of the liquid rate of any of wells—10 or 300 t/d, but first the specialist should get information about the well number, but still the specialist's estimate will be approximate.

Two explanations to the Table 2.1:

– For purposes of comparison, the amount of information for different parameters is assumed to have 128 values each. For liquid rates, this provides an accuracy of about 11 t/d, for water cut—about 0.005. In practice, this accuracy may be insufficient, and sometimes excessive, but in this case, only the main points are discussed, as the foundation for further presentation;
– With 128 possible values, there should have been 7 bits of information, but there are only 5.5 bits for the total number of wells and 2.41 bits for each well. This is because not all possible liquid rates are equally likely. The probability distribution of liquid rates of formation A shows that liquid rates up to 100 t/d are more likely. Which make up 75% of the total. Therefore, if this distribution is known, then this in itself gives some information about the parameter.

The bottom line is that if all the liquid rates were equally probable, and one had to predict the liquid rate at random, then no value would be better than the others would, but when one

Table 2.1 Liquid rate information capacity difference between formation A and wells

	Number of measurements	Minimum, t/d	Maximum, t/d	Mean, t/d	Standard deviation, t/d	Information capacity, bit	Portion of the maximum, %	Error information capacity, bit
Formation A	180,000	0.5	1400	82.7	111.2	5.5	79.1	5.4
Wells	25	4.2	563	79.6	25.1	2.41	34.4	1.46

knows that the most likely liquid rates are up to 100 t/d, then the forecasted value would be more likely to be correct. For the more concentrated distribution, it is easier to obtain a forecasted value. For example, a sample of reservoir pressure measurements for formation A contains only 4.3 bit of information (compared to 5.5 bits for sampling of liquid rates), so one can predict with an 85% probability that the pressure will be from 155 to 190 Atm. This is about a third of the total interval in which the reservoir pressure varies from 120 to 225 Atm.

If a scale of this kind is sufficient: 1—below average, 2—average, 3—above average, it is always possible to predict the expected value of reservoir pressure with a reliability of 85% by indicating the average division of this scale. This requires 1.06 bits of information if there is a probability distribution and 1.58 bits if there is none. Just as much information is contained in the phrase: "The reservoir pressure of formation A varies from 155 to 190 Atm", which can be found in geological reports. The phrase "The reservoir pressure of well 1001 is between 165 and 175 Atm" already contains 2.2 bits of information (3.32 bits without distribution).

Thus, we have come close to the question of predicting the value of a parameter and can state our main position: to predict the value of a parameter with a certain accuracy, one need to have a quantity of information corresponding to this accuracy. Moreover, this does not depend on the physical essence of the parameter, but is determined by the probability distribution that is available for this purpose. If there is no such distribution, then all parameters will require the same amount of information, depending on the accuracy with which one needs to know their values. Therefore, if a formula is proposed that

predicts the value of the parameter Y based on the specified value of another parameter X, then first it is necessary to analyze not the type of formula that connects these two parameters, but the amount of information that parameter X contains. No identical mathematical transformation adds anything to the amount of information that has been calculated according to the law of conservation of the amount of information. As a rule, the amount of information received at the output will be less than it was submitted at the input.

Consider the classic dependency: $Y = f(X)$, where Y—liquid rate, X—logging permeability. The amounts of information of these parameters for formation A are shown in Table 2.2.

It can be seen that the permeability sample contains two times less information than the liquid rate. This is due to a more compact distribution of the permeability, compressed by 73% into a single maximum, than for the liquid rate. Note that the porosity sample contains more information (5.5 bits) than the permeability sample. Both of these parameters are calculated from the same input logging parameter of alpha PS, so they are essentially the same value at different scales, but the calculation of permeability required more arithmetic operations, which was accompanied by a large loss of information.

It is obvious that the input information on the permeability is not enough to calculate the liquid rate, even if all this information was effective. However, if there is not enough information, then one can be satisfied with a rougher forecast. If the amount of information for both initial samples was determined under the condition of the same partition, the permeability was entered with an accuracy of 5 mDarcy, and the liquid rate—3 t/d. If one is satisfied with the accuracy

Parameter	Mean	Standard deviation	Information Capacity
Logging permeability	52.5 mDarcy	177.5 mDarcy	2.86 bits
Average liquid rate	81.2 t/d	93.2 t/d	6.03 bits

Table 2.2 Information capacity of input parameters

of the liquid rate of 6 t/d, the amount of information will still be insufficient, since in this case one needs 4.7 bits of information. Only with the accuracy of the forecasted liquid rate of 25 t/d, the amount of information will be almost enough—2.9 bits.

It is possible to use another way and to introduce permeability more fractionally, up to 2 mDarcy that will give 3.9 bits, but if judged realistically, it is impossible to take permeability with an accuracy of 5 mDarcy, as since the error of its definition is at least twice as high. This can be estimated approximately from the following considerations: the porosity and permeability contain the same amount of information, since they were calculated from the same input data, but the porosity varies from 15 to 25%, and even if one assumes that the accuracy of its determination is 0.1%, this allows to consider the accuracy of determining the permeability is not higher than 10 mDarcy, since its limits vary from 0 to 1800 mDarcy. Then the real amount of information in the sample of permeability is only 2.3 bits and this is not enough even for predicting the liquid rate even with an accuracy of 60 t/d. However, it is known from practice that there is no connection between the logging permeability and the liquid rates, and this is explained by the fact that the permeability contains insufficient information already at the input.

If we calculate the forecasted liquid rates using an exponential regression equation (it will be shown later that it is more adequate in this case than the polynomial equations): $Q = \exp(A * k + B)$, where A and B—regression coefficients, we will get only 1.13 bit of information for the sample of forecasted liquid rates, i.e. the efficiency coefficient was 39.4% (Table 2.3). However, with this amount of information at the output, it is impossible to predict the liquid rate no more than 200 t/d.

Visually, the result can be represented as shown in Fig. 2.2. This shows that a smooth exponential line (brown dots) is much more predictable than the chaotic spread of actual points (blue dots) and, therefore, contains less variety, and therefore less information, than the fact. In general, this is typical for most regression equations. As a rule, they are expressed as continuous functions of the polynomial type, while the original sample is discontinuous and almost the same permeability value is compared with completely different liquid rates.

This is all the information capacity of the forecast, which includes two stages:

1. Getting the equation of connection of two parameters, for example, such as shown in Fig. 2.2. It can be seen that there is indeed a connection between the logging permeability and the liquid rates, which is shown by a smooth curve;
2. Using the obtained dependence, calculate the forecasted liquid rate of the well. However, this is not possible in this case because there is not enough information at the input, and in addition, most of it is lost in the process of calculating the regression equation.

These two stages correspond to the solution of direct and inverse problems, and if the first of them has a single solution, then the second does not have a single solution at all, but rather a set of equally probable solutions.

It is almost impossible to predict precisely because the forecast method always assumes first obtaining a dependency based on the actual material and in this process, a large amount of actual information is always lost, and then this dependency is supposed to be used to accurately predict the facts. However, there is no longer enough information to do this, which is always lost in the process of generalization.

Table 2.3 Information capacity difference between input and output parameters

Type	Parameter	Mean	Standard deviation	Information capacity
Input	Logging permeability	52.5 mDarcy	177.5 mDarcy	2.86 bits
Input	Average liquid rate	81.2 t/d	93.2 t/d	6.03 bits
Output	Forecasted liquid rate	47.7 t/d	16.5 t/d	1.13 bits

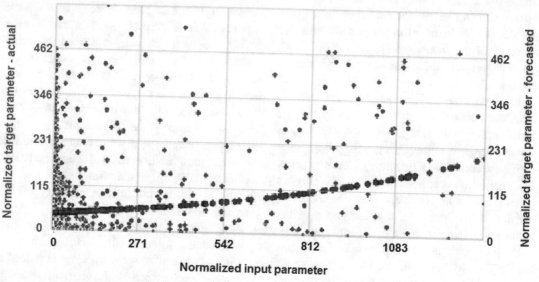

Fig. 2.2 Information capacity difference between actual and forecasted values

One can take several input parameters that add up to more input information, and therefore get more information at the output. However, losses begin immediately.

Let us assume that we will predict the liquid rate based on the product of the permeability by the net pay thickness of the reservoir (k * h), which is usually used in the hydrodynamic equations and in combination with the viscosity of the reservoir's fluid forms the main coefficient of the Darcy law: $q = -\frac{k}{\mu}\nabla p$, where q—liquid flux, ∇p—pressure drop over a given distance, and its integral form or, in other words, the Dupuit's equation: $Q = \frac{kh}{\mu}\Delta P / \ln\left(\frac{R_d}{r_w}\right)$. From these formulas, it is seen that in addition to the static parameters of the reservoir, the liquid rate depends on variable viscosity and pressure difference. It is known from practice that it is the quality of the formation that determines the liquid rate. Any pressure difference can be created

in low permeable formations, but the liquid rate will remain insignificant. Moreover, the calculation uses not the current, but the average flow rates of wells over their history, and they should be determined mainly by the properties of the formation. However, in this case, the interest is simply in the effect of multiplying the data.

First, the process of multiplying parameters leads to loss of information (Tables 2.4 and 2.5). Let us say we have two parameters, each of which takes a value from 1 to 8 with equal probabilities and, therefore, contains 3 bits of information. If we multiply them, we get a complex parameter containing not 6 bits that were submitted to the input, but only 4.789 bits or 79.8% of the input. This happened because, although the new parameter changes within a wider range: from −1 to 64, its values are not equally likely, so it contains less information. Note that if one increases the input parameters to

Table 2.4 Information capacity difference between initial and calculated parameters

Parameter	Mean	Standard deviation	Information capacity
Net pay thickness—h	8.56 m	2.71 m	5.69 bits
Porosity—Φ	20.4%	3.05%	6.12 bits
Permeability—k	17.9 mDarcy	69.6 mDarcy	3.56 bits
$\Phi * h$	176.5 % *mDarcy	67.05 %* mDarcy	6.34 bits
k * h	210.6 mDarcy* m	1004.06 mDarcy *m	3.14 bits

16, 32, and so on, the portion of output information will grow to 85% and further, but this is provided that the input parameters are distributed evenly. In reality, this does not happen, so the loss of information when multiplying is more significant.

For example, formation A shows that the product of permeability per net pay thickness contains even less information than the permeability itself, although an additional 5.69 bits of the thickness parameter were added at the input (although the same formation is specified, but the figures are slightly different from the above, since a different sample of wells was taken). Note that a simple calculation of the oil rate based on the liquid rate and water cut has a 52% information utilization rate.

First, it is noteworthy that the amount of input information was even less than just for the permeability, and the result is generally comparable except for a slightly larger standard deviation (by 20%) of the calculated liquid rate.

It is possible to show that other arithmetic operations on the parameters also lead to loss of information. If one replaces multiplication with addition in the above abstract example with multiplication of two parameters, the loss of information will even increase from 20.2 to 38.3%. The loss of information will increase even more when dividing and subtracting.

It follows that in relation to geological and field problems, there are hardly any formulas that can give a good result in the forecast. Sometimes it seems intuitively that the more complex the formula looks, the better the result it should give, but in fact, all this complexity leads to a loss of

information at every step of the calculation. The considered information approach makes it possible to evaluate the effectiveness of formulas based on the ratio of input and output information, without understanding their content.

It is even worse when information is used cyclically in the analysis process. For example, a method for calculating potential liquid rates is used to assess the prospects of a well for carrying out a work-over. Let us take a simplified version of the formula without gas and skin effect for clarity, since they do not change anything in essence, but only complicate the visibility of the calculation.

In a simplified form, the liquid rate of a well can be represented as $Qi = k * dPi/R$, where $R = \ln(R/r)$ is the logarithm of the ratio of the drainage radius to the wellbore radius, $k = kpr * h/m$—the coefficient of hydrodynamic conductivity: permeability multiplied by the net pay thickness of the formation and divided by the viscosity of the fluid. This value is determined by well testing, or it can be estimated as k * h according to the logging data, but as shown above, the information capacity of this parameter is small, $dPi = P - Pbh$—the difference of reservoir and bottom-hole pressures.

If one has an actual liquid rate and pressure difference, it is possible to estimate k: $k = Qi * R/dPi$, and then substitute the result in the same equation: $Q = (Qi * R/dPi) * dP\max/R$. The radius is reduced and it turns out that the potential liquid rate is proportional to the current one: $Q = Qi * dP\max/dPi$ or $Q = Qi * P/(P - Pbh)$. In this case, a forbidden technique is used, because the same information

Table 2.5 Information capacity difference between liquid rate values obtained by measurements and calculations

Parameter	Mean	Standard deviation	Information capacity
Permeability—k	52.5 mDarcy	177.4 mDarcy	2.86 bits
Net pay thickness—h	8.2 m	3.3 m	3.52 bits
k * h	520.2 mDarcy *m	1854 mDarcy *m	2.73 bits
Liquid rate measured	81.2 t/d	93.2 t/d	6.03 bits
Liquid rate calculated	47.9 t/d	19.9 t/d	1.13 bits

was used twice—first to get the coefficient of hydrodynamic conductivity, and then to calculate this information itself.

The use and significance of input information can be estimated by conducting a probabilistic simulation of the potential liquid rate based on the formation A statistics (Table 2.6).

The maximum and minimum in Table 2.6 are taken from actual data, and the average values are formed because of a complete search of all possible combinations, except for the explicitly forbidden ones—(P − Pbh) less than or equal to 0. As a result, the potential liquid rate contains only 19% of the input information $(3.77/(6.99 + 6.39 + 6.56))$, and the contribution of the actual pressure is a maximum of 38%, or $(4.24 * (4.24 + 6.99))$. Thus, the potential liquid rate calculated in this way is at least 62% determined by the current liquid rate of the well. In reality, there may be more, since the abstract iteration used, and the specific values of pressure and liquid rates have uneven definitions and the liquid rate has a larger information capacity, so its portion will increase. Even so, the potential liquid rate parameter does not have a larger information capacity than the average or maximum liquid rate of the well calculated over its history.

However, as a rule, there are no measured bottom hole pressures, and they are calculated from the liquid levels, as well as reservoir pressures using the formula: $P = f(Habs, Hdin, Poil, Pwtr, \%Wtr, Panul)$, which includes the absolute depth of the formation, liquid level of the well, oil density, water density, water cut, and annulus pressure of the well, respectively. Without going into the accuracy of the input parameters, we note that a number of them (Poil, Pwtr) is taken on average for the formation, therefore, it cannot contribute any information to the sample of reservoir (or bottom-hole) pressures. The absolute depth of the formation is a static parameter, i.e. all changes are made by the liquid level of the well and its water cut.

In Table 2.7, which shows statistical data of formation A, it can be seen that the information capacity of static and dynamic liquid levels is only slightly (by 5% and 10%. respectively) higher than the information capacity of reservoir and bottom-hole pressures.

According to the results of probabilistic modeling, the calculated reservoir and bottom-hole pressure contains 25–26% of the sum of input information, which gives 5–5.2 bits of information, provided that the water cut gives

Table 2.6 Information capacity assessment of potential liquid rate

Parameter	Minimum	Maximum	Mean	Standard deviation	Information capacity
Liquid rate	1 t/d	414 t/d	207 t/d	119 t/d	6.99 bits
Reservoir pressure—P	129 Atm	213 Atm	174.5 Atm	24.3 Atm	6.39 bits
Bottom hole pressure—Pbh	18 Atm	234 Atm	100.2 Atm	43.7 Atm	6.56 bits
P – Pbh	1.2 Atm	53 Atm	4.65 Atm	6.47 Atm	4.24 bits
Potential liquid rate	1 t/d	21,500 t/d	962 t/d	1545 t/d	3.77 bits

Table 2.7 Information capacity difference between pressures and liquid levels

Parameter	Mean	Standard deviation	Information capacity
Reservoir pressure	171.4 Atm	14.2 Atm	5.66 bits
Bottom hole pressure	125.6 Atm	35.7 Atm	5.98 bits
Liquid level static	205.0 m	205.4 m	5.93 bits
Liquid level dynamic	719.0 m	328.0 m	6.61 bits

5.73 bits and the absolute depths of the formation's top—4.3 bits. From this, we can conclude that the pressures recorded in the initial database have an inflated accuracy. In the process of probabilistic modeling, only 4.7 bits of information for the calculated pressures are obtained.

Thus, it becomes clear that when a hydrodynamic model is built, which contains millions of cells that still need to be multiplied by the number of months of the development history of the object under study, it would be good to first assess how much information is actually available for wells and how much of this information gets in the grid cell and on a specific calculation date.

Questions about the information capacity of geological and field data are closely related to uncertainties and errors of these data. The main problem is not the errors themselves, which could reduce the total amount of input information for modeling, but errors in the calculation results using inaccurate data. According to the error theory, the total error of the calculation result is determined by the formula: $O = \sum_{i=1}^{n} O_{pi} * \partial pi$. In other words, the error is equal to the sum of the products of errors in determining the O_{pi} parameter by the first derivative of this parameter—∂pi. Therefore, the main source of the total error is not inaccuracies in determining the geological and field parameters, but high values of the first derivatives in the functional dependencies. For example, the dependences of the liquid rate on the permeability or net pay thickness of the reservoir. In practice, it is often possible to see that small changes in the net pay thickness of 5–10% are accompanied by two to three times the change in liquid rates in neighboring wells. This shows that most of the geological and field tasks are incorrect.

Thus, when building geological and hydrodynamic models, it is necessary to take into account the amount of initial information and not to deceive anyone by building huge models from several hundred million cells, for which there is actually no necessary amount of information. All these considerations about the information content of the initial and calculated geological and field data were presented in order to show that the values associated with various parameters must be evaluated in terms of the their actual information capacity, and, most importantly, the actions performed with these values in the calculation process are always accompanied by information losses, sometimes more significant than the result. Everyone knows the aphorism of "garbage in, garbage out" of the computer, but still always hopes to get a good result somehow with the bad initial data.

References

1. Shannon, C. E. (1948). A mathematical theory of communication. *Bell System Technical Journal, 27*(3), 379–423.
2. Landauer, R. (1991, May). Information is physical. *Physics Today*, 23–29.
3. Rosinger, E. E. Mathematical model of information, ResearchGate.com website: https://www.researchgate.net/profile/Elemer_Elad_Rosinger/publications, Accessed on 21 April 2020.
4. Rosinger, E. E. Aspects of information in physics, ResearchGate.com website: https://www.researchgate.net/profile/Elemer_Elad_Rosinger/publications, Accessed on 21 April 2020.
5. Schweizer, D., Blum, P., & Butscher, C. (2017). Uncertainty assessment in 3-D geological models of increasing complexity. *Solid Earth, 8,* 515–530.
6. Wellmann, J. F., & Regenauer-Lieb, K. (2012). Uncertainties have a meaning: Information entropy as a quality measure for 3-D geological models. *Tectonophysics, 526–529,* 207–216.
7. Hansen, T. M. (2020). Entropy and information content of geostatistical models. *Mathematical Geoscience.* https://doi.org/10.1007/s11004-020-09876-z.
8. Ashby, R. W. (1956). *An introduction to cybernetics* (p. 295). NY: Wiley.

Contrasts Between Adaptive and Deterministic Models

In addition to the deterministic approach, there are other methods of geological and hydrodynamic modeling of petroleum reservoirs, which are no worse: "Some researchers are still convinced that the description of oil production processes can only be carried out on the basis of differential equations of the movement of liquids and gases in porous media and pipes. However, this approach does not reveal many essential properties of the reservoir. Like all large and complex systems, petroleum production facilities require the use of a whole hierarchy of models—from differential to integral, from deterministic to adaptive—that can describe not only different levels of system's organization, but also the interaction among those levels" [1].

The main problems of deterministic modeling are also listed in [1] and are as follows: "The lack of reliable information about the detailed geological structure of the reservoir and large errors in the field data. Thus, the accuracy of geological and petrophysical materials is so low that 3D geological and hydrodynamic models constructed using seismic data and logging permeability are no more than fiction (the accuracy of logging permeability determination is 100%). Under these conditions, integral 1D or 2D models are more accurate than 3D, since the integration errors are mutually compensated".

Deterministic models imply that all the geological and field data is obtained in clean laboratories of "quantum physics", and not on clay-filled drilling rigs or pads lost in deep forests, where the fittings of the wellheads grow like fancy iron trees. All this is well understood by supporters of alternative modeling methods: "But in the world of big data, strict accuracy is impossible, and sometimes undesirable. If we operate with data that is constantly changing, absolute accuracy takes a back front" [2].

The deterministic approach implies that there is a strictly defined connection between all objects and phenomena, and this connection can be unambiguously expressed through physical laws and differential equations. This direction comes from Descartes, who believed that everything in nature is rigidly defined and connected, and if one has complete information, one can accurately predict the future. It may be so. No one believes that events happen by accident. Randomness and probability are just a mechanism for describing events about which there is no complete information. This raises a very relevant question about whether it is possible to have complete information about anything, in particular about a petroleum reservoir. The deterministic approach implicitly implies that this is possible in principle, if not now, then in the future. The proposed adaptive approach assumes that this is not possible in principle—we will never be able to have complete information about a petroleum reservoir. This follows from the concept of "potential infinity" of information about a petroleum reservoir. In the sense that no matter how much information we have, new information can always be added to it.

S. Ursegov and A. Zakharian, *Adaptive Approach to Petroleum Reservoir Simulation*,
Advances in Oil and Gas Exploration & Production,
https://doi.org/10.1007/978-3-030-67474-8_3

Another problem is related to the fact that the mathematical apparatus of deterministic modeling is a priori considered the only correct one. It is represented by some finite set of functions (formulas). However, if we remember the fields of Hilbert's functions, we can understand that there can be no "only correct" functions. Any function is just one of many similar ones that give approximately the same result, but still in some ways different. This position is easy to check if one tries to find a regression equation to express the relationship between some parameters—there is a wide choice and no objective preferences. On the other hand, for example, there is a large set of equations, more than a hundred, for the so-called "liquid displacement curves". Moreover, one can make as many more without any possibility of finding any universal equation that would fit any petroleum reservoir. Having a set of such equations, one can choose the most appropriate curve for a particular field. Moreover, in this, one can see an element of adaptive modeling when a change in external conditions changes the calculation system; it adapts it to these conditions. The system in this case is a reservoir engineer. To go further and replace the engineer with artificial intelligence, we need to move away from a set of ready-made equations in the form of liquid displacement curves, and replace them with an artificial neural network or a cascade of fuzzy logic matrices. In this case, the mathematical apparatus will be formed based on actual data, will no longer be deterministic and will become adaptive.

The deterministic approach to modeling is so confident that it allows changing the initial data, for example, the reservoir permeability, so that it satisfies the deterministic approach equations, which remain unchanged. In contrast, the adaptive model never changes the initial data, but changes its mathematical apparatus so that it corresponds to these data. Despite the fact that the actual geological and field data is very far from true, but still they cannot be changed inside the modeling system. If this information is changed in the process of adjusting the hydrodynamic model to the fact, then there is nothing wrong with this. However, if one then uses this

corrected information for hydrodynamic calculations, it turns out that the same information is used repeatedly, which cannot be done. It is similar to trying to feed a cow by hay that has already been eaten a second time. Any algorithm can only use the information once. Now, if one changes the original data using some other algorithm that is not related to hydrodynamic calculations, and then submit the corrected data to the hydrodynamic calculation, it would be correct. Changing the data based on the results of the hydrodynamic calculation itself, as it were, determines in advance the result of this calculation and the forecasts that can be obtained using this calculation.

We can say that by allowing changes in the initial data, the deterministic approach violates the law of information conservation, expressed in the seventh theorem of Shannon. By correcting the initial data, the deterministic approach tries to increase the amount of initial information in the calculation using the calculation method itself, and this is not possible. The adaptive approach is free from this contradiction and this is its main advantage.

In the field of geological and hydrodynamic modeling of petroleum reservoirs, it is difficult to learn about the adaptive approach. The main stream of published scientific papers is devoted to the deterministic approach. However, adaptive modeling is popular in the economic sphere.

The concepts of adaptation and adaptability appeared, first, in the lexicon of economists with the arrival of system analysis in the economy. Almost all works devoted to the analysis of the properties of large systems of economy reveal the property of adaptability, that is, the ability to adapt; self-learning and self-organization. Thus, adaptation refers to the ability of a system to use the acquisition of new information to bring its behavior and structure closer to optimal under new conditions. Self-learning is the ability of a system to adapt to new conditions and correct its behavior based on mistakes made. The ability of the system to change its structure, composition and parameters of elements when changing the conditions of interaction with the environment is distinguished as a property of self-organization.

Any large system is adaptive—it adapts in one way or another to changed conditions. However, not each of these systems has the property of self-learning-adaptation not only based on external information, but also because of how far the behavior of the system is far from optimal. "The highest level of survivability of a large system is determined by the fact that it has not only the properties of adaptability and self-learning, but also self-organization" [3].

However, in the economic sphere, adaptation is sometimes understood in a limited way as a reconfiguration of time series prediction functions (smoothing, autocorrelation) or retraining of neural networks as new information becomes available. At the same time, the mathematical apparatus and the system itself do not change in essence. However, if the neural network is retrained and its structure remains the same, then there is no adaptation. This is understood by economists themselves: "It seems difficult to draw a clear line separating adaptive forecasting methods from non-adaptive ones. Already forecasting by the method of extrapolation of ordinary regression curves contains some element of adaptation, when with each new acquisition of actual data, the parameters of the regression curves are recalculated and refined" [1]. True adaptation could occur when the structure of the neural network changed.

The principle of adaptability was first formulated by Ashby. "In his theory two cycles are necessary for a system to be adaptive: The first feedback loop operates frequently and makes small corrections. The second feedback loop operates infrequently and changes the structure of the system, when the "essential variables" go outside the bounds required for survival" [4]. However, Ashby worked in the field of psychiatry, where the true meaning of adaptation was obvious to everyone. A human being adapts to new conditions of life, while changing the stereotypes of their behavior.

The proposed geological and hydrodynamic models are called adaptive in order to distinguish them from traditional deterministic models. The meaning of the proposed model's adaptability is not that it is updated as new data becomes available, but that it adapts to the conditions of each new modeling object. Besides, the method of calculating deterministic models varies for different petroleum reservoirs. To do this, these models have a wide range of tools, for example, one can choose a different method for interpolating petrophysical data or boundary conditions on wells. In this sense, a simulation engineer is a "second feedback loop", but this is not a property of the deterministic model itself.

The proposed adaptive model adjusts itself to each object, since all its algorithms are based on the principle of self-organization. These are either machine learning methods or cellular automata or lattice Boltzmann methods. The simulation engineer does not participate in their selection at all. The engineer's role is generally limited to the preparation of initial data. Only if the modeling object does not have a unified database from which to get the necessary information, so one should have to take this information from different sources, structure and correct it. As a rule, this is done manually Most often it is necessary to check and correct stratigraphic markers. However, then the entire calculation of the adaptive model is fully automated in one stream: the geological model—the hydrodynamic model—the forecast of oil production levels—the forecast of well work-over results. The simulation engineer cannot influence the course of this calculation in any way.

Special attention should be paid to the question of the number of layers of the 3D geological model. When using deterministic simulators, this number is set by the simulation engineer. In the proposed adaptive model, the number of layers is determined by the structure of the petroleum reservoir. Here, the simulator adapts to the initial data, not the other way around. The complexity of an adaptive model is determined by the amount of initial information. This model does not contain any hard-defined equations or functions that can be written analytically as implemented in deterministic models. In this way, deterministic models seem to show the superiority of their mathematical apparatus over the initial information. The device is considered correct and if the data contradicts it, then they need to be corrected. Certainly, developers of deterministic models understand the ambitiousness of this statement and constantly improve the mathematical apparatus of their models as new modeling objects and new situations appear, but this mechanism of the deterministic model is not intended to adapt to new initial information every time.

Let us now consider the question of dimensions. Geological and field information is diverse in its subject content and in the dimensions that display this content. For example, the coordinates of wells, the reservoir's depth and its gross and net pay thickness are measured in meters, porosity and oil saturation-in fractions of units, permeability—in mDarcy, oil and water production—in tons, reservoir pressure—in atmospheres. The values of these parameters are generally not comparable to each other and for a simulation engineer they all make sense because the value of any parameter creates a substantive representation of it. In the adaptive model, all geological and field parameters are normalized using a hyperbolic tangent in the range from -1 to $+1$. After that, they are indistinguishable from the subject point of view. They only contain information in the same quantities as before. The normalized values retain their diversity. For a simulation engineer, these figures will no longer cause objective representations. However, for the mathematical apparatus of the adaptive model, these representations are not necessary, it only needs information. That is why all the geological and field parameters are normalized in order to get rid of objectivity and move to pure information, to its diversity. This is the main difference between the proposed adaptive and deterministic models.

Further, the mathematical apparatus of the adaptive model immanently takes into account all the uncertainty and inaccuracy of the geological and field parameters. Since the parameters loaded into the model do not have absolute values, only their relative values are relevant: where greater, where less. The simplest illustration of this position of the adaptive model looks like this —let us assume that the oil saturation and, accordingly, the phase permeability of the model

cell for oil are equal to zero. Then filtration of oil through such a cell in a deterministic model is impossible, since this model strictly observes physical laws. However, in the adaptive model, oil filtration through such a cell is possible, because the adaptive model is able to correct the critical boundary of filtration conditions, i.e. adapt to them. This is possible because the adaptive model understands that all the values included in the filtration equation are highly unreliable and therefore should not strictly control the behavior of the reservoir system. Indeed, the authors of deterministic models also understand this, but this is not provided for in the mathematical apparatus of their models, which is rigidly bound to physical laws.

The understanding that not everything can be squeezed into the paradigms of classical science exists among unbiased scientists: «Under closer observation, it has become evident that natural phenomena do not behave as they are though subject to the narrow determinism postulated by the paradigm of classical science» [5].

In the adaptive model, negative values of oil saturation in the reservoir are allowed. Certainly, the real oil saturation cannot be less than zero. However, the mathematical meaning of the oil saturation parameter is that the smaller it is, the lower the phase permeability and the less likely it is that oil will flow through the cell. If the oil saturation value is negative, it means that the probability of such a flow is the lowest. However, it cannot be zero; because the model cell has a large size, for example, $2500\ m^2$, if the model grid step is 50×50 m. Each square meter of a cell can have its own oil saturation, and the oil saturation of the cell is output as an average. Its value means that at some points in the cell, the oil saturation can be quite high and therefore the overall probability of oil flowing through this cell is not zero. Negative oil saturation and negative density of oil reserves occur if a well takes more oil than there are reserves in its area, even if it is large enough. In the case of a deterministic model, they try to correct its geological basis to provide the necessary reserves. However, this is not quite correct because the geological model was built in a deterministic simulator with all available data. If it is corrected, there will be no gain from the information point of view. The resulting contradiction between the geological and filed data contained its own information, which was destroyed in the process of adjusting the geological basis of the deterministic model.

The hydrodynamic model and the distribution of remaining oil reserves obtained using it are necessary, first of all, in order to predict the further oil production of a particular well. In the deterministic model, this is done using the Dupuit's equation with different additional coefficients, and therefore, at zero value of the phase permeability, there will be zero oil production. However, the adaptive model builds a predictive fuzzy logic function using machine learning, which allows the use of zero phase permeability, because in fact there was oil production. For this calculation, absolute values are not important, but only relative values. Thus, the adaptive model takes into account the fact of uncertainty and inaccuracy of the available geological and field data, which, however, are not adjusted for the mathematical apparatus of the model and retain all their information, which is available even in their inconsistency.

It is appropriate to quote the following statement by Ashby: "Cybernetics treats not things but ways of behaving. It does not ask "what is this thing?" but "what does it do?"… It is thus essentially functional and behavioristic. Cybernetics deals with all forms of behavior as far as they are regular, determinate, or reproducible. The materiality is irrelevant… The truths of cybernetics are not conditional on their being derived from some other branch of science. Cybernetics has its own foundations" [6]. The meaning of this statement is that cybernetics is not interested in the physical essence of things and phenomena, but only their behavior, which is described by the laws of information and, first of all, the law of necessary diversity. It also follows and in the following, a more modern statement: "People are used to looking for reasons in everything, even if it is not so easy or useful to establish them. On the other hand, in the world of big data, we no longer have to cling to causality. Instead, we can find correlations between data

that provide us with invaluable knowledge. Correlations cannot tell us exactly why an event is happening, but they do warn us about what kind of event it is. In most cases, this is quite enough" [2].

If we return to the example of permeability, then cybernetics is not interested in the physical nature and its diversity, i.e. in some cells of the model permeability more and no less, without going into a study of the reasons why the distribution was exactly that. Most of the filtration flow will be directed to cells with higher permeability, but there is no question about its absolute value. Because this value cannot be determined accurately.

The initial position of deterministic modeling was to reproduce the actions of simulation engineers in the computer. However, keeping in mind the statement of A. Turing that a machine thinks differently than a human being, the construction of the proposed adaptive model is conducted exactly as the computer understands it and precisely for its needs. The difference between a simulation engineer and a computer, first, is that the specialist works visually, looks at maps and cross sections, and 3D visualization is well developed in deterministic models. It is like a computer game. When people look at beautiful images, it seems that they have a deep content. At some stage, companies that produce computer equipment and software tried to convince everyone that it is necessary to consider 3D geological and hydrodynamic models on huge monitors, that this allows the specialist to enter the reservoir, but it seems that this idea has already outlived its usefulness. It would make sense to look at the topographical map of a petroleum reservoir like this, if it were possible, but not its model, which has already greatly smoothed and distorted reality and in fact the specialist looks at what he or she built, so the specialist cannot see anything interesting.

Because the computer works only with numbers, then the visualization for the adaptive simulation is not of great importance. The model has a value only in the numbers that can be removed from its cells. This raises the question of whether the logic of such a process is correct, when the actual information, for example, from a thousand wells is sprayed into a million cells of the model, and then collected at the points of wells in order to predict their behavior. Whether it is possible to make an immediate forecast based on actual data without using a model. Indeed, there are machine learning experts who believe that the geological and hydrodynamic models are not necessary at all and it is enough to simply use the correlation dependencies of the initial data. They say so precisely for reasons of information. As a rule, such specialists use artificial neural networks for their modeling.

In this regard, it is necessary to pay attention to the structure of artificial neural networks with deep learning, which is characterized by the presence of more cores in the intermediate layer than in the input layer. For example, a perceptron with three inputs. Data from these three inputs is divided into five intermediate layer cores, and then the outputs from these five cores are summed in a single output core. From the point of view of information theory, the most interesting point here is that there are more cores in the intermediate layer than there are inputs. However, it is obvious that each core of the intermediate layer has less information than one input. Information is sprayed across the cores of the intermediate layer. The main purpose of this design is to increase the number of weight coefficients of the artificial neural network. The more of them, the more complex the dependence that the neural network displays and the less predictable the forecast result, i.e., the more the artificial neural network looks like a "black box".

In some respects, the proposed adaptive model resembles a huge neural network, although it is not in the literal sense of the word. Existing wells can be considered as inputs. Their geological and field data are sprayed onto the cores of the intermediate layer, i.e. the model cells. In the process of calculating the model, it is like setting up weight coefficients, for example, the density of remaining oil reserves in the model cells or the current pressure field.

Further, as it happens in some artificial neural networks, the weight coefficients are modified by the interaction of the cores. Then the outputs of the cores or cells of the model are summed. In fact, an adaptive model is an implicit analog of an artificial neural network. Not literal, because there are no such networks in this model, but in a global sense.

Note that artificial neural networks usually have a simple structure with only a few tens or hundreds of weight coefficients. Therefore, their result is generally quite clearly predictable and sometimes close to the regression polynomial. For example, a third-degree cubic polynomial with three inputs has twenty coefficients, and a simple perceptron-type neural network with three inputs and five cores in the intermediate layer has twenty-nine coefficients, so their results are quite close. Artificial neural networks differ from living neurons in their simplicity. Let us say a living neuron has one output (axon), but it has a million fibers, and it is possible that different signals go through them.

In order to display a complex physical system, one needs an adequately complex model. A petroleum reservoir is a very complex system, and to display it requires a complex model of millions of cells, but there is too little very noisy and contradictory initial information for them. This is why deterministic modeling methods cannot be used, since they implicitly perceive the available information as true by their mathematical apparatus and impose strict restrictions on the model. This is how they determine the forecast result in advance. It is due to the lack of initial information that the geological and hydrodynamic model should be sufficiently free in its behavior. Then the forecast will not be predetermined, but will be unexpected and useful.

Deterministic simulators have great capabilities and can be configured for anything, but why. The proposed adaptive model is somewhat similar to unpredictable artificial intelligence. It works on the principle of self-organization like cellular automata, which are widely used in this model. It has a relatively small number of invariant restrictions, for example, the flow of a liquid goes from a point with a high pressure to a point with a lower pressure. Then the flow is free in its behavior. Such things are sometimes called "black boxes" and this is understood in a negative sense. Because if it is not possible to trace all the steps of the calculation, then such a computing system is not suitable. However, not everything can be traced and proved. This is what the Godel's first incompleteness theorem states: "There are unprovable truths in any system" [7].

References

1. Mirzadzhanzade, A., Khasanov, M., & Bakhtizin, R. (1994). *Modelling of complicated systems of oil extracting etudes. Nonlinearity, nonuniformity, heterogeneity* (pp. 464). Ufa: Gilem.
2. Mayer-Schonberger, V., & Cukier, K. (2013). *Big data: a revolution that will transform how we live, work and think* (pp. 242). Canada: Eamon Dolan/Houghton Mifflin Harcourt.
3. Lukashin, Y. P. (1989). An adaptive method of regression analysis. In P. Hackle (Ed.), *Statistical analysis and forecasting of economic structural change* (Ch. 13, pp. 209–216). IIASA, Springer-Verlag.
4. Ashby, W. R. (1960). *Design for a brain: The origin of adaptive behavior* (p. 286). London: Chapman and Hall.
5. Bale L. S. (1995) Gregory Bateson, cybernetics, and the social behavioral sciences. *Cybernetics & human knowing, 3*(1), 27–45.
6. Ashby, R. W. (1956). *An introduction to cybernetics* (p. 295). NY: Wiley.
7. Gödel, K. (1931). Über formal unentscheidbare Sätze der Principia Mathematica und verwandter Systeme I. *Monatshefte für Mathematik und Physik, 38*(1), 173–198.

4

Alternatives for Mathematical Apparatus of Adaptive Simulation— Neural Networks and Fuzzy Logic Functions

Today, in the era of computer domination, computational mathematics has changed. If earlier one had to find analytical expressions or formulas for fast calculations, now this is not necessary. Any formula is just a sequence of actions with variables. A similar sequence of actions can be set in a computer using a set of subroutines, and most importantly, one can find a solution that differs from the usual analytical expressions, but gives approximately the same result.

The main mathematical method of the proposed adaptive model is machine learning and working with big data sets. In fact, this approach has a statistical basis, although the statistical apparatus itself is used very sparsely. Mainly in the framework of the mean, variance, shape of distribution and linear correlation coefficient. Methods of nonparametric statistics and numerical solutions of differential equations are also used. However, the main methods are fuzzy logic and cellular automata or lattice Boltzmann methods.

All calculations are based on Kolmogorov's theorem that any complex function can be represented as a superposition of a finite number of primes [1]. Any algorithm is simple in itself, and complexity occurs when hundreds or sometimes thousands of such algorithms are superposed. The main idea of the mathematical apparatus of the proposed adaptive model is that the petroleum reservoir and the processes occurring in it are too complex to describe them by any regular functions. Figuratively, this can be represented as follows: the drawing is a strict geometric shape, and the artist's picture is not, but the picture more accurately describes reality.

Accounting for uncertainty

There is a purely ideal idea that you first need to check or "verify" all the original data and only then build your own model. This is not the case, because the source data can generally only be checked by statistical methods, comparing them with each other. We are not talking about mechanical errors, but rather about the uncertainty of the initial data. However, if one uses mutual verification to delete the data with the most errors, the same errors will be found in the remaining data that previously seemed correct. Therefore, there is no point in checking and rejecting anything, but you need to take the original data as it is, while understanding that the result will also be characterized by uncertainty. Nevertheless, this uncertainty should be understood as a rough scale of measurements, in the sense that it is impossible to get a model with less uncertainty than its initial data.

The tasks of analyzing and forecasting the state of petroleum reservoir development are mostly incorrect. A small change in the argument (input parameter), for example, the net pay thickness may correspond to a large jump in the function (target parameter), for example, the average well rate for oil. In this regard, the results

S. Ursegov and A. Zakharian, *Adaptive Approach to Petroleum Reservoir Simulation*, Advances in Oil and Gas Exploration & Production, https://doi.org/10.1007/978-3-030-67474-8_4

of any calculations may contain large errors. As it was said in Chap. 2, the calculation error is equal to the sum of input parameter errors multiplied by their partial derivatives. The errors themselves can have different signs and can be destroyed mutually if there are a large number of parameters, but the partial derivatives in the case of an incorrect problem can be very large, so the final calculation error can also be significant. In this regard, it is impossible to predict accurately any particular results, for example, the rates of individual wells. However, on average, the forecast will be effective, because some errors are eliminated mutually when summing the results of forecasts for individual wells.

Normalization of parameters

For multi-parameter calculations, it is not possible to use absolute parameter values, since they have significantly different dimensions. These values must be normalized in the range either from 0 to 1, or from −1 to 1. At the same time, the distributions of geological and field parameters are often close to the lognormal, which is extremely inconvenient, since a significant part of the interval does not actually work. One can pre-log the parameters and then normalize them. In any case, the issue of parameter normalization is one of the most important in the calculation of the proposed adaptive model.

One should also keep in mind that any statistical method averages the forecast results, drawing them to the maximums of distributions. As a rule, the variance of forecast values is significantly lower than the variance of the initial data. The adaptive model is calculated using its own normalization method, which results in a bipolar distribution (Fig. 4.1).

Cumulative cascade of neural networks

The mathematical apparatus of the proposed adaptive model also includes the artificial neural networks. The main problems of artificial neural networks are the inability to formalize the choice of their structure and their retraining. These problems were solved using a cumulative cascade (Fig. 4.2). It consists of a set of identical simple perceptrons with three inputs and one output. There are five nodes in the intermediate layer. In general, this perceptron resembles a transistor and is designed to amplify the signal. The correlation between the calculated and actual values of the target parameter in such a perceptron is always higher than the correlation of any of the three input parameters with the target. At the same time, the structure of such a perceptron is simple, and it cannot be retrained.

The cumulative cascade is formed in such a way that the number of perceptrons in each layer is a multiple of three. For example, as shown in Fig. 4.2, there are twenty-seven simple perceptrons in the first input layer, nine in the second, three in the third, and only one in the last and fourth. The outputs of the three perceptrons in the first layer are set to one of the perceptrons in the second layer, the outputs of the three perceptrons in the second layer—one of the perceptrons of the third layer, etc. Figure 4.2 shows a four-layer cascade with five layers, the first layer will have eighty-one perceptrons, with six layers, two hundred and forty-three, and so on. A large number of input parameters can be applied to the input of such a cascade, and some of them, the most significant ones, can be applied two or three times in different triples of the first layer.

As shown in Fig. 4.2, weak correlations of input parameters with the target, for example, well rate for oil, the cumulative cascade significantly strengthens. However, this may not be enough. Then the cascade can be clustered. If its structure is preserved, there will be seven perceptrons at each position instead of one, as shown in Fig. 4.3. They are arranged in the form of an inverted pyramid, opposite to the pyramid of the cumulative cascade itself. The entire training sample is fed to the first perceptron, at the top of the pyramid. Then analyzed setup errors for each line in the sample, and they are divided into two parts, for example, one will be the error with a negative sign and the other with positive. The next two perceptrons are trained and then the sample of each of them is again divided into two parts, following the error sign. This results in four samples for training four perceptrons at the base of a pyramid of seven perceptrons. If the seven perceptrons of the

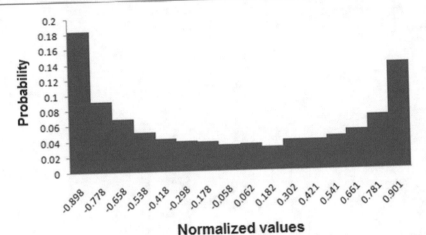

Fig. 4.1 Distribution of a normalized parameter

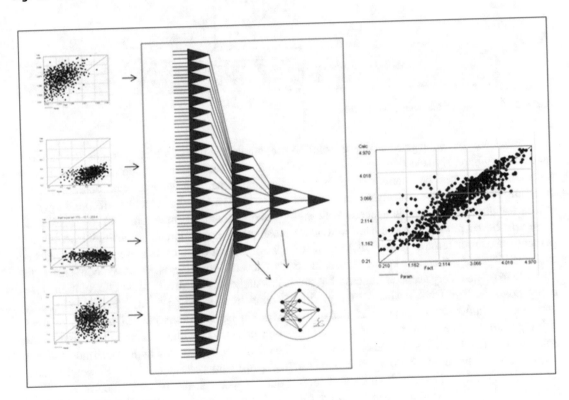

Fig. 4.2 Cumulative cascade

pyramid are numbered from 0 to 6. Then for each well, for each position in the cumulative cascade, you can choose a number that will allow the calculated value at this position to have the smallest error. By building numbers in a chain, one can get something like DNA for each well. Figure 4.3 shows the DNA for two wells and shows that they are different. In a five-layer cascade with eighty-one perceptrons, the first layer has a chain of 121 digits, and as practice has shown, these chains never coincide in any pair of wells. The total number of perceptrons in a five-layer clustered cascade will be 847.

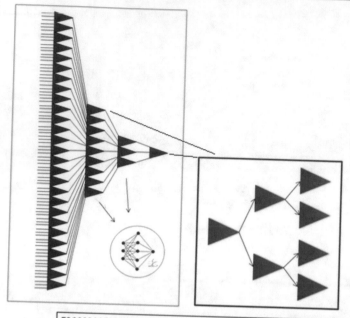

526662665222655252206266526565565221000255565005525062266555652626526552526566552555
5205420653006144623651326425544015220203133662265632232500522655250552356054665521064

Fig. 4.3 Clustering of the cumulative cascade

A clustered cascade gives a dense solution space everywhere. By selecting the DNA strands for each well, it can be configured so that the correlation between the calculated and actual values for the training sample exceeds 95% and there will be no retrain, because each individual perceptron cannot retrain. If the DNA chain of each well in the forecast was known, it would be possible to make a very accurate forecast, but this is not possible. The DNA chains for the predictive sample almost never match the chains of the training sample. The only way to set a certain number of DNA chains for each well, for example, using the Monte Carlo method, and as a result get several hundred or thousands of forecast values for it and build a probability curve for them, which is more correct than just taking any one value.

Fuzzy logic functions

Any function that connects two variables can be set analytically, for example, as a polynomial or using an artificial neural network. It can also be set in its natural form, i.e. in the form of two data columns, the first of which specifies an argument, and the second—a function. Indeed, this is cumbersome, the sample can contain several thousand rows or more and correspondingly the same amount of numbers and by setting an analytical expression or an artificial neural network, the set of thousands of numbers is reduced to several coefficients. This is convenient for calculations. However, this convolution loses some of the original information. Certainly, when it comes to samples of millions of rows, one cannot do without convolution, but in the field of geological and hydrodynamic modeling of petroleum reservoirs, there is not so much data, so you can use natural functions.

If one sorts a selection of two parameters by the first column, i.e. by argument, then the graph of the function will take the form shown in Fig. 4.4a. This shows that there is a correlation between the argument and the function, but the values of the function fluctuate very much, and if one sets a small interval for the values of the argument at each point, then one can observe a wide variation in the values of the function

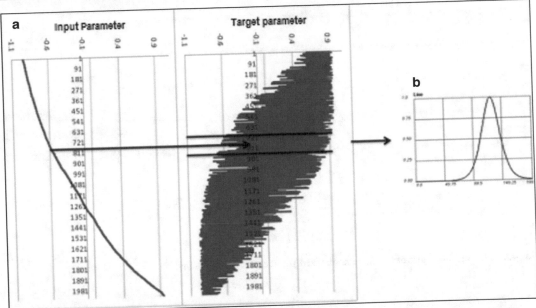

Fig. 4.4 Behavior of the target parameter in relation to the input parameter, **a** sorted functions of input and target parameters, **b** membership function of the target parameter on the selected interval

within this interval. For the selected interval, one can calculate the average value and variance, and build the membership function as shown in Fig. 4.4b.

This membership function can be built for each argument value. If there are several input parameters (arguments) for a single target parameter (function), for example, three as shown in Fig. 4.5, then each of them will have its own membership function. These functions can be projected onto an adder (a straight-line segment that covers all possible values of the function). For normalized parameters, this is the interval from −1 to +1 (in Fig. 4.5—along the X axis). As a result, we get the classic version of summation of membership functions. The Y axis in Fig. 4.5 shows the weight value for each possible value of the membership function, the calculated value of which is selected based on the maximum total weight (red line). This is an elementary calculation option.

In practice, everything is more complicated and it turns out not so unambiguously (Fig. 4.6). This shows that the total line can have several maxima, but it is easy to find the global maximum (by the largest value).

In practice, in a given range of argument values, one can iterate through all the set of pairs of argument and function values and construct a different membership function for each of these pairs (Fig. 4.6). Then it becomes clear that the calculated value of the membership function is possible for the entire interval, but with different weights, and as which the global maximum is selected. This calculation mechanism is used in all cases of using many parametric fuzzy logic functions in the proposed adaptive model. These functions are written to binary files and used for various purposes.

Let us assume that each pair of input and target parameters sorted by input parameter is an elementary fuzzy logic function. In an adaptive model, there may be several hundred thousand such elementary functions.

Next, it should be taken into account that any elementary fuzzy logic function can be divided into two parts, which will have correlations of opposite signs (Fig. 4.7). If we use one elementary fuzzy logic function for the entire sample, we get an averaging of these two opposite trends. In this case, one can build two elementary fuzzy logic functions for each sub-sample and perform

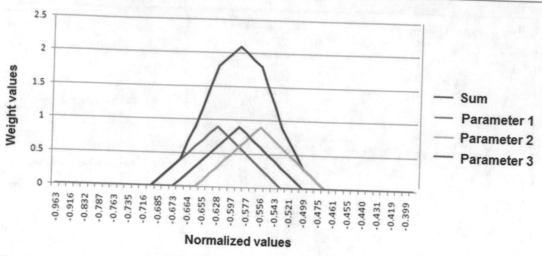

Fig. 4.5 Membership functions of three normalized parameters and their sum

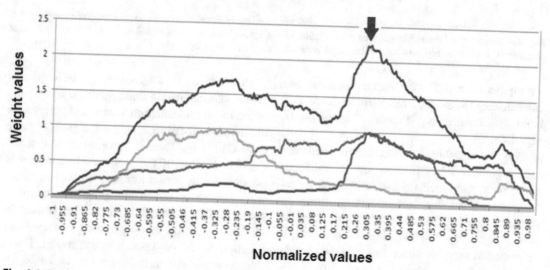

Fig. 4.6 Results of three membership functions summation

calculations on both, and then select the priority one.

If there are several input parameters, it is possible to iterate through all or part of the combinations of these parameters with different correlation signs and get a variation of the calculated values of the function.

Fuzzy logic matrix

If one forms a function with two input parameters (arguments) in the form of a square matrix, along the axes of which the normalized

dimensions of these input parameters are deferred, namely from −1 to +1, then the matrices will have a square shape. The number of nodes on the X and Y axes is the same and can vary from 50 to 500 depending on the number of rows in the training sample. As a result, the matrix will have from 2500 to 250,000 nodes. The more of them there are, the more detailed the matrix will work. The coordinates of each node are determined by the normalized values of the input parameters. The value in the node corresponds to the sum of the target parameter values. Since several target

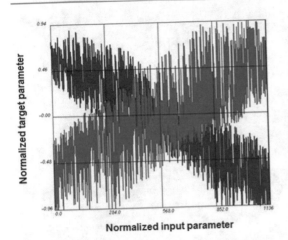

Fig. 4.7 Two opposite trends in the same sample

parameter values fall into the node, you can calculate the average value and variance for it, and then use them to build the membership function. Accordingly, two matrices are constructed for each pair of parameters, one for the average values, and the second for the variances. The matrix view is shown in Fig. 4.8.

In principle, it may turn out that not all cells of the matrix are filled, because there is not enough volume of the training sample. Then they can be filled in using a self-organizing method that resembles cellular automata. Note that the matrix grid is hexagonal for a more symmetrical operation of the mechanism.

Fig. 4.8 View of a fuzzy logic matrix

When calculating the forecast for any pair of input parameter values, one can get the mean (v) and standard deviation (dv) of the target parameter. Using these two values one can construct a membership function of the form $y = 1.0 - \tanh(((x - v)/dv) * f)^2$, which will be defined in the range from $v - dv$ to $v + dv$. At point v, the value of the function is +1, at the borders of the interval—0. The coefficient f determines the shape of the distribution within the borders of the interval.

Cascade of fuzzy logic matrices

A single matrix contains only two input parameters, and this is not enough to solve complex problems. Therefore, a cascade of fuzzy logic matrices is necessary. Let us say there are four input parameters (A, B, C, D), from which one can make six pairs (AB, AC, AD, BC, BD, CD). Note that all parameters are normalized in the range from −1 to + 1, so the coordinates of the sides on all matrices are the same. Two matrices are formed for each pair: the first one is for the average values, and the second one is for the variances. As can be seen from Fig. 4.9, all the matrices look different.

There are various methods of working with multidimensional space; the proposed adaptive model uses the method of fuzzy logic matrices. It is a classical scheme of analysis: first, the space is decomposed into component components, the regularities of each of them are identified, and then the synthesis or integration of these components is performed, because of which a forecast is obtained. The membership functions are summed up in the same way as already described.

A cascade of fuzzy logic matrices works much like an artificial neural network. The results of their calculation are correlated with each other by 70%. However, unlike a cascade of fuzzy logic matrices, an artificial neural network is not able to predict something new. Its main feature is image recognition. When making a forecast, an artificial neural network can only show what was in its training sample, and it finds the closest option for this. A cascade of fuzzy logic matrices,

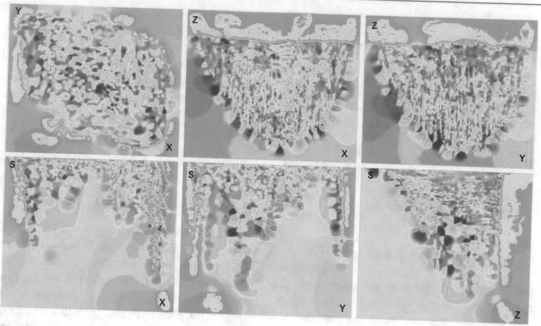

Fig. 4.9 Cascade of six fuzzy logic matrices of four input parameters (X, Y, Z and S) and porosity as the target parameter

on the other hand, can show the forecast result, i.e. a new result that was not present in the training sample due to random summation of membership functions.

We cannot say that this cascade already works much better than artificial neural networks, but it contains much more coefficients. If the artificial neural network has only a few tens or hundreds of coefficients, then the cascade has hundreds of thousands or millions. The number of coefficients depends on the number of input parameters of the cascade. For example, with nine parameters, there are thirty-six pairs of matrices, which gives 9000 coefficients for matrices of 2500 nodes. Indeed, such a huge cascade requires a lot of memory and time for calculations, but it is more like the work of a living brain, which stores matrices of images from millions of elements.

Mechanisms of cellular automata

Cellular automata most correspond to the paradigm of the proposed adaptive model, since they clearly manifest self-organization. They are widely used, including in fuzzy logic matrices, through which most of the functions used are expressed.

The adaptive model uses three mechanisms of cellular automata. The first version of the mechanism, which is close to the classic one, is as follows. First, the matrix is filled with the actual values of a parameter defined in wells, such as porosity, at points defined by a pair of coordinates of this matrix. At the same time, the standard deviation matrix is initialized at the same points. Certainly, these points are significantly fewer than the nodes in the matrix, and you need to fill in all the nodes. This is done using the lattice Boltzmann method. Nodes that contain the actual values are fixed and these values do not change in the future. All other nodes are assigned values randomly, according to the actual distribution. To do this, a row with the source data is selected randomly and its value is assigned to an empty cell. The result is a chaotic picture (Fig. 4.10a). At the same time, the conjugate standard deviation matrix is also transformed in the same way. Next, an optimization mechanism based on the idea of the lattice Boltzmann method is launched. The values of neighboring cells are exchanged so that the differences between the values in neighboring cells are generally minimized. At the same time, the values of cells in the standard deviation matrix change. However, the cells where the actual data was stored remain unchanged. This allows the process to converge in 2000–3000 iterations. One also needs a certain trend surface that affects the coefficients that control the exchange of values between neighboring cells. As such a trend, one can use seismic data that is defined for all nodes of the matrix. As a result, the matrix shows a pattern of distribution of the parameter relative to its coordinates (Fig. 4.10b).

The second, more trivial mechanism used by cellular automata resembles interpolation. It is used for calculating fields in the geological model. First, one also sets the parameter values at the well points. Then they begin to "spread" around the surrounding cells. In this case, not only the parameter itself is transferred to neighboring cells, but also the weight, which decreases as the number of transitions between cells increases. In principle, each cell only works with neighboring cells and "does not know" how far it is from the actual point. In this version of the mechanism of cellular automata, it is allowed to use a nonlinear weight function, which decreases with each step. This requires three auxiliary empty matrices. On the first auxiliary matrix, the weight decreases linearly, for example, the weight of the cell that received data is 0.999 of the weight of the cell that transmits this data. This is a linear relationship. On the second auxiliary matrix, the coefficient is equal to the weight of the transmitting cell multiplied by the coefficient of the first auxiliary matrix. This is a non-linear relationship. In the third auxiliary matrix, the coefficient is equal to the weight of the transmitting cell multiplied by the coefficient of the second auxiliary matrix. This is a higher-order nonlinearity (Fig. 4.11).

The third mechanism of cellular automata is mainly used in the calculation of the hydrodynamic model. This mechanism is closer to the lattice Boltzmann methods, since weight functions in real numbers are used to determine transitions between neighboring cells. In this case, the rule is observed that transitions can only be between neighboring cells. In the case of a hydrodynamic model, these are transitions in the amount of oil and water. First, one sets the volume of selected oil in the cells through which the

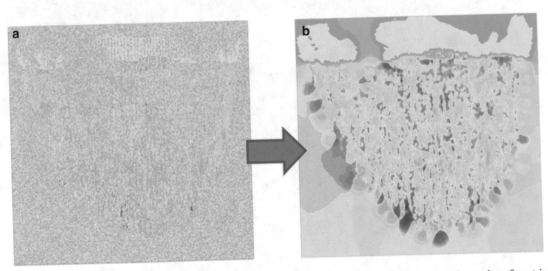

Fig. 4.10 Example of the mechanism of cellular automata, **a** results of matrix initialization, **b** results of matrix calculation

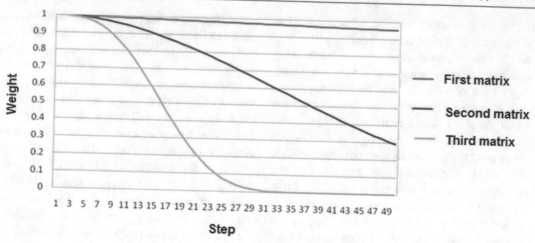

Fig. 4.11 Illustration of a non-linear decrease of weights of fuzzy-logic matrix according to the steps

wells pass, and then these volumes begin to "spread" throughout the model polygon. At the same time, there is no weight function that decreases at each step. The cell does not "remember" the past at all. The transitions are influenced by "guiding" fields, such as the pressure and phase permeability fields.

The main prospects for further development of the mathematical apparatus of adaptive modeling are related to the expansion and development of functions based on cellular automata, since they are the least predictable and therefore interesting.

Reference

1. Kolmogorov, A. (1950). *Foundations of the theory of probability* (p. 80). New York: Chelsea.

Any geological model is essentially a geometric representation of a petroleum reservoir, and it would seem that there should be no big differences between the deterministic and the proposed adaptive version of this model. However, these differences are also, above all, in the overall thickness and number of layers. The deterministic models are developed extensively in the direction of thinning layers to 0.4 m and increasing the number of cells to hundreds of millions. This contributes to the development of used computers, increasing their RAM and acceleration of video cards, but in principle, it is closer to computer games than to artificial intelligence. Everything is aimed at improving visualization, so that one can view the created geological model from different angles.

Something similar is done in archaeology, for example, to model the inner galleries of the Egyptian pyramids, although there is more initial information for this purpose in the form of thousands of photographs of these galleries. In the visualization there is a meaning, it can help to see something interesting, which was not noticed before. However, in fact, all this visualization turns out to be false, since the model cells are interpolated along a sparse grid of wells. To build such a cumbersome model, there is no necessary amount of initial information, and this model only seems detailed. Certainly, in addition to wells, one can attract seismic data, but its vertical resolution is much rougher than 0.4 m.

Slicing layers is the weakest side of a deterministic geological model. Petroleum reservoirs just cut into layers, starting from the bottom, in two ways, either the thickness of the layers is smooth and there is not enough total thickness of the reservoir layers come out, or the thickness of the layers is proportional to the thickness of the layer and then the layer thicknesses can be reduced to the millimeter. From the point of view of dividing space into cells, everything seems to be correct, but the studied geological space is anisotropic, and the slicing of layers in the deterministic model completely ignores this. Strictly speaking, each layer extends over the entire area of the model and it should be understood that it is correlated everywhere and thus the layers set a false structure of the reservoir space. It is known from practice that it is impossible to perform a detailed correlation of layers with a step of 0.4 m. After all, the reservoir usually has a lenticular structure, even between neighboring wells, some lenses can pinch out, and some appear.

An even more controversial stage in the construction of a deterministic geological model is upscaling. There is no reasonable explanation for why one cannot immediately build a grid of the hydrodynamic model for wells, but it must be enlarged from the grid of the geological model. Maybe someone wants to say that there was a detailed geological model, and now it needs to be coarsened in order to get a hydrodynamic model. However, the detail of the model is determined not by the size of the grid, but by the amount of input information on wells and seismic surveys. The other way, justified by human practice, seems more reasonable—to go from the general to the particular—first to build a rough hydrodynamic grid, and then to refine it to a geological one. As is customary with artists, first, a sketch is made with general masks, and then the details are worked out. Since a detailed geological model is only needed for visualization, the comparison with artists is quite justified. The proposed adaptive model does not have upscaling. An adaptive geological model is built not for visualization, but for the needs of the system itself, as the basis of a hydrodynamic model and for machine learning for subsequent forecasting [1–4].

One of the main differences between the adaptive geological model and the deterministic one is that the adaptive model in some cases does not coincide with the well data. This seems unacceptable, but the model is not needed to coincide with wells, but to forecast the geological structure of a petroleum reservoir in its undeveloped zones. The adaptive model is a function. Let us consider for example some regression function, for example, a polynomial of the first-second degree. Its calculated values also do not match the original points and this is normal. If we take a higher-degree polynomial, we can achieve that its calculated values completely coincide with the actual ones, but this polynomial cannot be used for forecasting.

Now let us look at the adaptive geological model. It does not coincide with wells in cases where closely located wells have significantly different values of the interpolated parameter. This happens due to measurement errors, especially well survey, if the wells are inclined. Let us say a well passes through a cell and a well parameter is assigned to this cell, for example, the absolute depth of the reservoir top. This cell has an area of $2,500 \text{ m}^2$ and its edge zones, which are located from the well at a distance of 25–30 m, are affected by data from the neighboring well, which is located at a distance of 150–200 m. Let the influence of the neighboring wells are only 20%, but it can be significant to

the cell assigned the weighted parameter different from a parameter value crosses this cell in the well. One can discard some of the wells that show the greatest differences from the model, which is often done, or correct the parameters of these wells to make them closer to the model. However, in both cases, the initial information will simply be lost and nothing will be gained in return, except the smoothness of the drawing. If we approach the model as a function, then variations in the initial parameters in the wells do not interfere with anything. The model seems to average and smooth out well data.

The second significant difference between an adaptive geological model and a deterministic one is that it has significantly fewer layers. As a rule, the number of layers in an adaptive model does not exceed six or eight. In this case, the average layer thickness is 5–10 m, in rare cases more. The point is that the adaptive model does not try to present itself as continuous, which is achieved by the deterministic model, thinning its layers, but is clearly discrete. A petroleum reservoir can be divided into very small parts, up to individual grains of sand. In any case, the model cell has a much larger volume than a grain of sand and the parameters assigned to it are always averaged. From this, the larger the cell volume, the more stable the average value of its parameters. If the layer thickness in the deterministic model is reduced to 0.4 m, the cell area is still large and does not correspond to this thickness. When the layer thickness is 5–10 m, it corresponds to the size of the cell horizontally. In the adaptive model, even for large petroleum reservoirs, its grid is set in increments of no more than 25 m.

When a reservoir is divided into layers, it is assumed that each layer represents a genetically unified sedimentation cycle, with its own patterns. This can later be used for machine learning to distinguish one layer from another due to these patterns. It would be good if the model layers were selected manually using detailed correlation methods. This would increase the amount of input to the model. Nevertheless, such a detailed correlation is too time-consuming and almost never occurs, especially on large deposits. However, in an adaptive geological model, layers of equal thickness are never cut.

The meaning of the detailed correlation used in the adaptive model is not to track the point-to-point genetic features of the formation of a petroleum reservoir, but to correlate the extent of permeable layers. One would like that the layer was characterized by permeable sub-layers connected, but separated from adjacent layers by non-permeable sub-layers. Thus, each layer was a quasi-single separate reservoir. Therefore, the adaptive model cannot set an arbitrary number of layers, but only as many as can be selected from the geological data.

The adaptive model divides into layers as follows. Select those wells for which there are stratigraphic markers and selected permeable sub-layers. These wells are sorted by the long axis of the petroleum reservoir, for example, by the Y-axis, since the reservoirs are more likely to have a sub meridional orientation. The selected layers are normalized in such a way that they are divided into layers 1 m thick and they have both permeable sub-layers and non-permeable ones, which are no less important. In addition, the selected layers are aligned to the bottom of the reservoir and instead of absolute depths on the borders of the layers, the distances from the bottom are indicated.

We start with the southernmost well and look for the first permeable sub-layers from the bottom. Then we look for it in the neighboring wells, focusing on the distance from the bottom. If one can find it there, then continue the search from the following wells as long as the search does not degenerate. This determines the permeable lens of a certain length, which was opened by a certain number of wells. Some lenses break off immediately, and some are found in hundreds of wells. In this process, adjustments are made for changes in the layer thickness and for the fact that the layers may be slightly displaced relative to the bottom, even in neighboring wells.

Then go back to the first well and look for the second permeable sub-layers from the bottom. If all the layers are exhausted in this well, then go to the next well and start from it. Repeat this process until all permeable sub-layers are assigned to some lenses, each of which has its own number. Tens of thousands of such lenses can be isolated in a large petroleum reservoir. Indeed, the entire process is automated and performed by a computer. Having finished with the permeable sub-layers, we do the same with the non-permeable ones. Then we combine the results into a single table or display them on a graph (Fig. 5.1).

In Fig. 5.1, the X-axis shows the distance from the bottom of the reservoir, and the Y-axis shows the number of wells that opened a particular lens. The red line corresponds to permeable layers, and the blue line corresponds to impermeable ones. Choose the maximum on the blue line closest to the bottom of the reservoir, but so that between it and the bottom there is a maximum of permeable sub-layers. Draw the cross section along the blue maximum and "cut" the lower layer from the reservoir. At the next stage, one needs to build a map to make sure that the resulting layer is stable. Sometimes one has to go back and choose a different cross section.

After "cutting" the first layer, repeat the process and cut the second, then the third, and so on until all the layers within the studied reservoir are "cut". How many of them will turn out depends on the structure of this reservoir. The layers should not be too thin, as this will further increase the number of active cells in the hydrodynamic model and increase the calculation time. Thus, we get detailed markers for a multilayer geological model of the studied reservoir. All layers have different thickness, which depends primarily on the structure of the reservoir. At the end, the system smoothes the layers a little in order to eliminate outliers, but keep the logic of the thickness of all layers. To divide the gross thickness of the studied reservoir into 6–8 layers, it takes one and a half to two hours. However, this is done once by the computer when forming the project.

Reconstruction of petrophysical parameters

As a rule, some permeable sub-layers do not have their petrophysical parameters—porosity, permeability, or oil saturation, although the gross

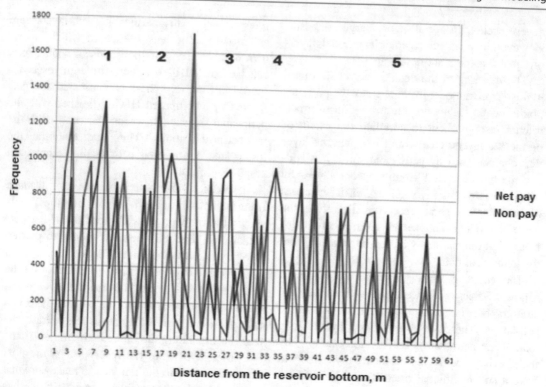

Fig. 5.1 Statistics of permeable sub-layers distribution inside a separate layer of the adaptive geological model

thickness of these layers is known. It is usually accepted to interpolate any parameter from the actual data and exclude from consideration those wells where this parameter is not defined. However, one can go the other way and first calculate the petrophysical parameters in all the intervals where permeable sub-layers are selected, and then interpolate. This path is preferable because more input information can be used for this calculation.

The calculation is performed by using a cascade of fuzzy-logic matrices based on current values of five petrophysical parameters: the absolute depth of the top of the permeable sub-layer, the distance from the middle of the sub-layer to the top of the reservoir, the distance to the non-permeable sub-layer above, the distance to non-permeable sub-layer below. In addition, two parameters that reflect the distribution structure of permeable sub-layers in the well in relation to the structures of neighboring wells are also calculated (Fig. 5.2). To do so, the net-to-gross ratio is used. Indeed, it does not change in

the well, but it creates a difference between wells, which is reflected in the values of the calculated parameters.

This calculation is based on the values of petrophysical parameters that are defined in the wells. It is also possible to use available core data, primarily for calculating permeability and logging parameters and well rates for oil and liquid. There is not always a complete set of the listed input parameters, but the system uses those available for each well.

Addendum sub-models

The geological space of a petroleum reservoir is purely anisotropic, because it is formed in layers over millions of years. During each year, the sedimentation conditions also change, so the reservoir consists of thin layers that can be seen by studying the core. In this regard, it makes no sense to use 3D interpolation when calculating the geological model. This interpolation is not used in the proposed adaptive model.

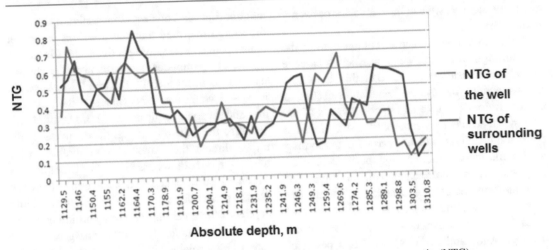

Fig. 5.2 Parameters of distribution of permeable sub-layers share in wells-net to gross ratio (NTG)

After the layers are selected, a separate geological model is created for each of these layers, and then they are added together into a general model. To do this, addendum projects are created, geological and field data related to this layer are copied there, and calculations of all addendum projects are run simultaneously. To save time, these calculations are parallelized. This logic of layer model calculations is separately justified by the fact that the layers were distinguished as single genetic formations, with their own internal patterns, which can be traced in changes in the general thickness or net-to-gross ratio. Also because machine learning methods are used for calculating layer models in addition to simple interpolation.

Base fields of sub-models

In our opinion, interpolation is an incorrect method, although it is not yet possible to get rid of it completely. The problem is that any cell in the model is interpolated only if it falls inside a triangle with three wells at its vertices. If the cell is located outside of this triangle, then one needs to go to extrapolation, which is not very reliable. If the relative interpolation errors are 15–30% of the calculated value, which can be determined by excluding each well in turn and calculating its parameter for the rest, the extra errors immediately increase to 100–150%.

When calculating geological models of layers, the maximum use is made of seismic data that is defined for all cells. The calculation requires at least three structural surfaces based on seismic data. If these data are not available, then they have to be calculated from wells, which is highly undesirable. After all, seismic data contains much more information than well data.

It does not matter which reflecting horizons the structural surfaces are built on. It is desirable that they are as close as possible to the studied reservoir. The set of structural surfaces reflects the history of the reservoir's development. The distances between them show the rates of sedimentation. By themselves, they show tectonic activity at different times. It is known from practice that the foundation surface is always informative. If there are three structural surfaces, then one can get three variation fields of these surfaces from them. They reflect the curvature and possible activity of post-sedimentation processes that affect the formation of fracturing and petrophysical properties. Then there are two thickness fields between these three surfaces. This results in eight parameters. Two fields of total derivatives along the X and Y axes are added to them, as well as a terrain surface that displays neo-tectonic movements. This creates a vector of eleven parameters for each model cell. If there are more than three structural surfaces available, say six, we can form a vector of twenty parameters.

Next, a training sample is compiled from the cells through which the wells pass, and a cascade of fuzzy logic matrices is trained on this sample. At the same time, the system has some freedom of choice. It does not take all the components of the vector, but selects only nine of them based on the correlation analysis, which is carried out according to the minimax principle—the parameters must have the greatest correlation with the target, for example, with the gross thickness, and the smallest between them. Nine parameters allow you to create a cascade of matrices from 36 paired layers (average value and variance). This is a powerful system that contains an average of 180,000 coefficients, and is able to display the complexity of parameter distribution within the model.

These matrices are compiled for the five main parameters of the geological model: gross thickness, net-to-gross ratio, porosity, permeability, and oil saturation. They are used for calculating the base fields of these parameters. From Fig. 5.3, it follows that the matrix quite adequately reflects the patterns of distribution of the gross thickness, revealing details that are not directly related to the wells. Figure 5.4 also shows that for the permeability logarithm, one can get a rather interesting distribution field.

In principle, one could just use these fields for geological modeling, but they are slightly refined due to well data. To do this, data from these wells are entered in the cells through which the wells passed, and then data from the base field is entered further along the 150 m grid, whose nodes are located at distances of at least 150 m from the wells. After that, interpolation is performed using a method that resembles cellular automata.

To do this, one needs to have a trend that will guide the movement of cells (Fig. 5.5). It can be obtained from seismic parameters, or calculated from source data, but it is better that the trend is not obtained from the same data. In addition, three grids are created, along which the weights move. At the first stage, the cells through which the wells pass are initiated. These cells are given an initial weight equal to one. The active cells in which the parameters are initiated pass it to the neighboring six cells and thus activate them. At the same time, they transmit a weight parameter whose value is already less than one. Moreover, the weight parameter decreases non-linearly and its value is influenced by the trend. More weight is passed when the trend values in the transmitting and receiving cells are close. If the difference is large, the transmitted weight decreases. This rather directs the flow. It is preferable to go where the conditions are close. For example, if a structural surface is taken as a trend, then the flow is preferably to the point where the absolute depths of the receiving and transmitting cells are close. For example, if a map of the gross thickness is interpolated, it can be assumed that the conditions of sedimentation on the flank and in the dome of the structure were different. Therefore, the original active cells located on the flank will spread along it, and not move towards the dome of the structure.

A cell that passed its parameter to a neighboring cell loses its activity. One cannot pass anything to it or change its parameter in any way. However, the cell that the parameter and weight was passed to becomes active. In principle, two or three neighboring cells can pass the parameter to it, and then the value in this cell is defined as the weighted average. Its weight is determined similarly. The process continues until all cells in the model get their parameter and weight. At first glance, this looks like a normal inversion, but it is not. In inversion, the calculated cell "knows" at what distance it is from the wells. In our case, it "doesn't know". It only contacts its immediate neighbors, as it should in cellular automata. As a result, we get a more free and predictable distribution of the parameter. This method also has prospects for amplification, which is not present in the usual inversion.

Note that initially activated well cells do not change in the future during a single calculation. This is too strict a condition. In order to reduce this rigidity of cell initialization, the sample of wells is sorted by parameter value and divided into three or four parts. Further interpolation is performed separately for each part of the sample. At the same time, interpolated values are entered in the places of missing wells, which will differ

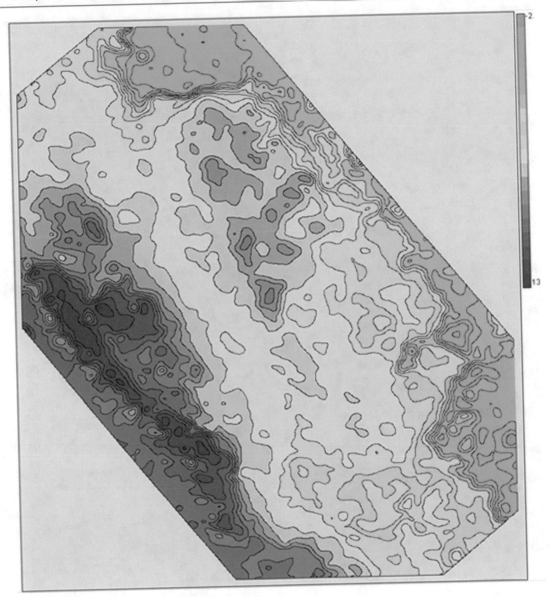

Fig. 5.3 Base map for the layer gross thickness

from the parameters of these wells. In this case, interpolation does not cover the entire area of the model, but only the first 15–20 steps, so that the effect is only on the cells of neighboring wells. Then all four calculations are added together. This gives the model some freedom from downhole data, which in principle may be erroneous and it makes no sense to adhere to them too rigidly. As a result, the model never exactly matches the wells, but it more objectively displays the distribution patterns of geological parameters.

Having the main maps on them, one can calculate the fields of parameters necessary for the hydrodynamic model: pore volume per square meter, hydrodynamic conductivity—permeability multiplied by the net pay thickness and density of original oil reserves per square meter. The hydrodynamic conductivity parameter is normalized in the range from 0 to 1, since there is no

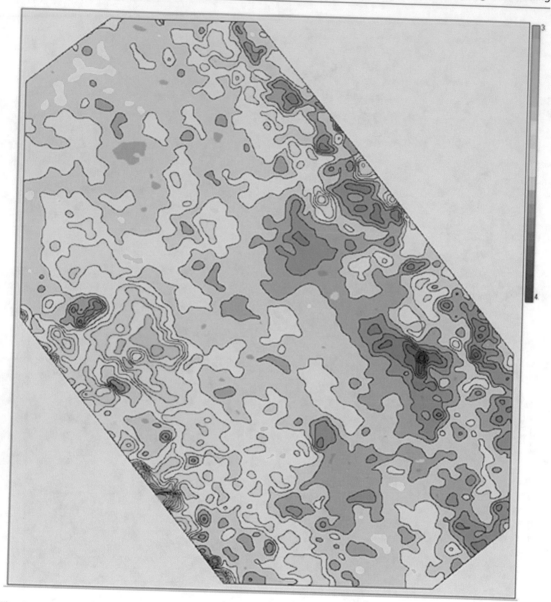

Fig. 5.4 Base map for the layer logarithm of permeability

need for its absolute value, but only relative. However, the hydrodynamic conductivity calculated only from petrophysical data is not sufficient. Therefore, it is refined due to two additional fields constructed according to field parameters—the liquid rates of producers and the injection rates of injectors, which are more informative than petrophysical parameters. The first of them is the field of well interaction

coefficients, which are calculated from the data of time series of liquid rates and injection rates. There are three such calculations.

The first calculation is related to the estimation of the interaction of producing and injection wells. To do this, take one of the injection wells and several neighboring producing wells, within a radius of about five hundred meters. Using the time series of liquid rates and injection rates,

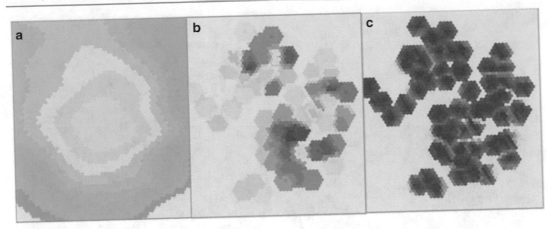

Fig. 5.5 Diagrams of the interpolation process, **a** trend, **b** parameters, **c** weights

a fuzzy-logical multi-parametric function is built, in which the injection rate of the injector depends on the liquid rates of neighboring producers. Certainly, from a physical point of view it should be the opposite and that the liquid rates of producers should depend on the injection rate of injector, but nothing prevents to turn this feature because the injection rate increases the growth of the repression on the reservoir, and with a large liquid production in the vicinity of the injection well is reduced reservoir pressure and thus increases repression. Having built such a function, we begin to "disable" the production wells in turn to assess the change in injection rate. In this case, the calculation is stable because the injection well is not compared with one producing well, but with several.

Then the coefficients of mutual influence between producing wells are calculated in a similar way. To do this, a function is constructed in which the liquid rate of the producing well depends on the liquid rates of the neighboring producing wells (Fig. 5.6). This results in a second coefficient.

However, wells occupy only a part of the model area and in order to fill the rest of the area, approximately the same calculation is performed as for petrophysical data. To do this, we take a series of permeable—impermeable sub-layers in the well and make up approximately the same function as the distribution of permeable sub-layers in one well, depending on the distribution of permeable sub-layers in neighboring ones. Neighboring wells are also excluded and the interaction coefficient or correlation coefficient of well sections is estimated. This is the third coefficient that allows to cover the entire area of the model. Looking ahead, we note that at the time of this calculation, there is a distribution of permeable sub-layers not only on the actual wells, but also on additional nodes that are located on the entire area of the model in increments of 150 m and at the same time no closer than 150 m to the actual wells.

Now having the weight coefficients, one can put them in the middle of the distances between pairs of wells and build a field of mutual influences. Where wells are located, the total coefficient is calculated mainly from them, and where they are not—from petrophysical data. The coefficients are normalized in the range from 0 to 1. Then one can multiply the hydrodynamic conductivity field by them and make it more differentiated.

The next step is related to the construction of a field of logarithms of the maximum liquid rates of producing wells (Fig. 5.7). This is done using a cascade of fuzzy logic matrices for nine input parameters, but in this case these are not parameters taken from structural maps of seismic exploration, but parameters of the geological model of the layer already built at this time: gross and net pay thicknesses, porosity, permeability, oil saturation, pore volume, reserve density, etc.

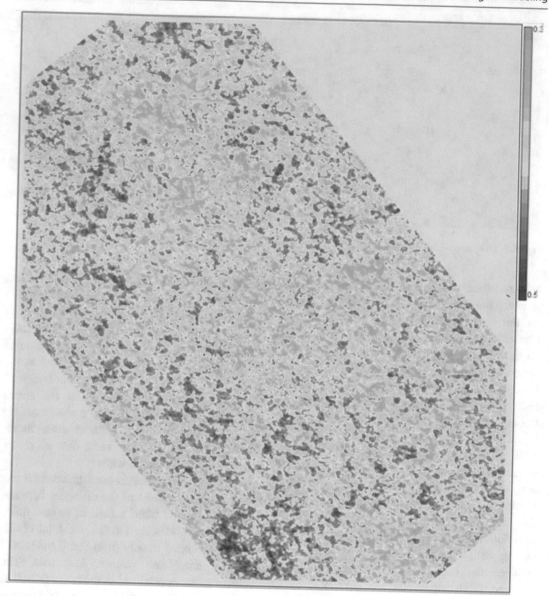

Fig. 5.6 Map of the coefficients of mutual influence of the wells

To do this, a part of the well's liquid rate is taken that relates to a specific layer. This portion is calculated in the main multi-layer model when all the geological models of layers are already available based on the petrophysical parameters of these layers. Having obtained the field of maximum liquid rates, the field of hydrodynamic conductivity is modified once again, and in the end it turns out to be more meaningful than if it was calculated only from petrophysical data. Here, as it were, the geologic model is adjusted to the history, which when using a deterministic approach is done at the stage of hydrodynamic modeling, when the permeability is changed in order to get a given rate of the liquid. The adaptive approach immediately takes into account the rates of the liquid when calculating the hydrodynamic conductivity field.

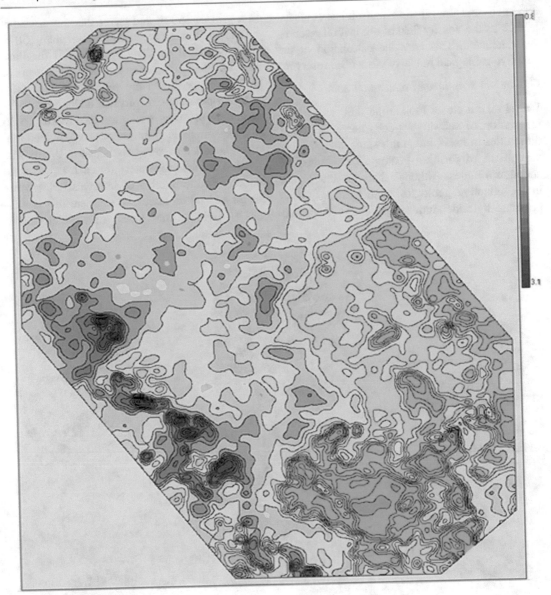

Fig. 5.7 Map of the logarithm of maximum liquid rates

General scheme of the adaptive geological model calculation

The geological model is calculated in five stages (Fig. 5.8). At the first stage, data is prepared, stratigraphic markers are corrected if necessary, and the petroleum reservoir is divided into layers. The second stage calculates addendum sub-models based on the main five parameters. In the third stage, these models are combined into multi-layer ones, the geological and petrophysical sections of addendum nodes are calculated, oil and liquid production, as well as injection is divided into layers and all this is distributed across sub-models. At the fourth stage, the sub-models calculate the fields of interaction between wells and the maximum rates of the liquid. At the fifth stage, these fields are integrated in a multi-layer model, the hydrodynamic conductivity field is corrected, and the oil-water and oil-gas contacts, as well as the density field of original oil reserves are calculated. This ends the calculation of the adaptive geological model of the studied reservoir.

Geological structure of addendum nodes

Let us take a closer look at the third stage. The integration of sub-models does not present any difficulties in itself, but at this stage, there is also a problem of forming geological structures of addendum nodes. Although there are few layers in the adaptive geological model, this is compensated by addendum nodes. They are located on a grid of 150 m and cover the entire area of the model, and are located no closer than 150 m from the actual or planning wells. At the third stage of calculation, when the main parameters of the multilayer model are ready, the geological and petrophysical structures of addendum nodes (including planning wells) are calculated. To do this, the structures of these nodes are divided into intervals with an average thickness of 0.4 m. The division is made separately for each layer of the model, and the number of intervals is the same for all addendum nodes. In essence, this is similar to dividing into layers of a deterministic

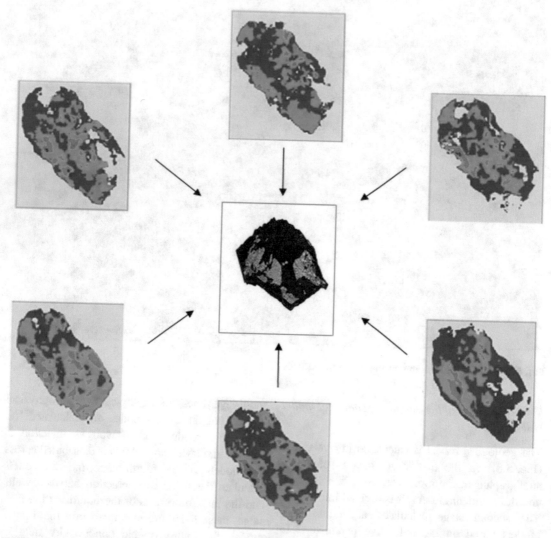

Fig. 5.8 Integration of sub-models into a multi-layer model

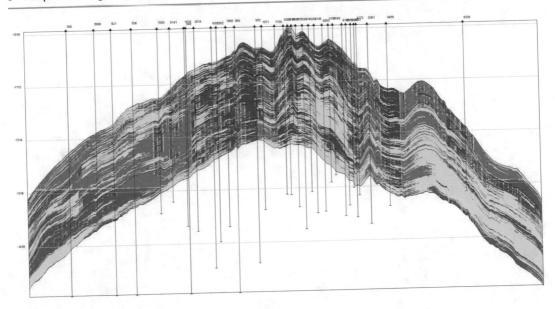

Fig. 5.9 Geological cross-section

geological model, but with a relatively large grid step. In addition, there is no overall bulky thin-layer grid. Node data is stored in binary files, from which it is downloaded if necessary. This is necessary when drawing a cross-section through the reservoir, which can be obtained in real time simply by drawing the line of this cross-section on the map (Fig. 5.9).

The structures of addendum nodes are calculated by interpolation using the mechanism of cellular automata. To do this, first create the same structures for the actual wells, and then interpolate them. The most difficult point here is associated with the allocation of intervals of permeable and impermeable sub-layers. Interpolation is made on the model layers with a thickness of 0.4 m. For each such sub-layer, the probability that it is a permeable sub-interlayer is calculated. Further, based on the known net pay thickness of the layer at the point of any addendum node, one can make this thickness from the sub-layers with the highest probability that it is a permeable sub-layer. Sections of addendum nodes are then passed to the sub-models and used for calculating mutual influence coefficients.

The grid of the adaptive geological model is directly used for the hydrodynamic model without any upscaling. Only non-permeable cells are removed from the grid of the hydrodynamic model in order to reduce the consumption of RAM, which is usually very significant for hydrodynamic calculations.

References

1. Ursegov, S., Zakharian, A., & Taraskin E. (2018). A new adaptive identification and predicting system of sweet spots and their production potential in unconventional reservoirs. In *Proceedings of the 20th EGU General Assembly Conference*, Vienna, Austria (p. 15594).
2. Ursegov, S., Zakharian, A., & Taraskin, E. (2018). Full field unconventional carbonate reservoir modelling using adaptive simulation technique. In *Proceedings of the 80th EAGE Conference and Exhibition 2018: Opportunities Presented by the Energy Transition*, Copenhagen, Denmark.
3. Ursegov, S., Zakharian, A., & Serkova, V. (2018). Geological modeling of reservoir systems—An adaptive concept. In *Proceedings of the 16th European Conference on the Mathematics of Oil Recovery*, Barselona, Spain.
4. Taraskin, E. N., Zakharian, A .Z., & Ursegov, S. O. (2018). Implementation of adaptive geological modeling for supervising development of the Permian-Carboniferous reservoir of the Usinsk field. *Neftyanoe khozyaystvo—Oil Industry*, *10*, 36–41.

Adaptive Hydrodynamic Modeling

It is the method of hydrodynamic calculation that most distinguishes the adaptive model from the deterministic one. Deterministic hydrodynamic modeling is supposedly based on differential equations, in particular the continuity equation or mass conservation, which in the Cartesian coordinate system is written as:

$$-\left[\frac{\partial(\rho u_x)}{\partial x} + \frac{\partial(\rho u_y)}{\partial y} + \frac{\partial(\rho u_z)}{\partial z}\right] = \frac{\partial}{\partial t}(\emptyset\rho) + \tilde{q},$$

where ρ—fluid density, Φ—porosity, u—fluid velocity, $\tilde{q}$—sink or source of mass.

However, in the best-known book on deterministic hydrodynamic modeling, Aziz and Settari [1] said that from a practical standpoint it is hopeless at this time to try to apply these basic laws directly to the problems of flow in porous media. Instead, a semi empirical approach is used where the Darcy's law is employed instead of the momentum equation. The Darcy's law, experimentally established in the century before last, is much easier to write down:

$$u = -\frac{k_{pr}}{\mu}(\nabla\mathrm{P} + \rho g),$$

where k_{pr}—absolute permeability, μ—fluid viscosity, ∇P—difference between reservoir pressure and well bottom hole pressure.

It has been experimentally established that in multiphase filtration, the Darcy's law can be widely considered valid for each phase separately. However, all these experiments were conducted under surface conditions with relatively high speeds corresponding to everyday life while maintaining the Newton's laws. Hydrodynamic processes in the underground geological space proceed according to completely different laws. For comparison, let us say that in quantum mechanics, Newton's laws are not fulfilled, because there are too high speeds and small times. We can assume that in geologic space-time, Newton's laws are also not fulfilled, but this is due to too low speeds and long times, as well as due to thermobaric conditions.

The biggest problem of the deterministic approach is the discretization of time and space in the transition from differential equations to finite-difference ones. As it is said in [2]: "In general, it is not possible to prove the convergence of finite-difference filtration equations to differential ones. This is due, first, to the strong non-linearity of the equations, especially in terms of conductivity, and second, to the need to determine the effective characteristics for each calculation block, due to the heterogeneous structure of the layers".

In fact, the discretization is very rough. The volume of a hydrodynamic model cell can reach 1000–3000 m³ and the reservoir space in this volume is highly differentiated, and a single permeability parameter is assigned to the entire cell, as if the cell is a point in space with an infinitesimal volume. All variables related to differential equations must be continuous. This

S. Ursegov and A. Zakharian, *Adaptive Approach to Petroleum Reservoir Simulation*,
Advances in Oil and Gas Exploration & Production,
https://doi.org/10.1007/978-3-030-67474-8_6

means that if there is a point, it can be arbitrarily approached by an infinitesimal distance. However, what is a point in geologic space? Approximately, we can say that this is the average distance between wells, which reaches 100–300 m. This is the size of the geological point. Accordingly, the geological space in the mathematical approximation cannot be continuous.

In deterministic modeling, it is assumed that shredding the grid will improve the case. However, this is a purely formal approach, since a crushed grid will not have more variety than a large one, since this is determined by the source information. A fine grid cannot significantly change the flow distribution.

The second point of continuity is time. Usually the calculation is done in increments of one month and this is very rough, if one remembers about differential equations. The simulator understands this and, if the process does not converge, begins to split the time step of the calculation, which of course slows it down a lot. Despite the fact that the time steps become small, the debit remains the same, since it is recorded in the source database in increments of one month. Ultimately, we can say that deterministic hydrodynamic modeling is a method of material balance on a finite number of cells.

Methods of mathematical analysis are very limited in their application to the description of the geological structure and development of a petroleum reservoir precisely because of the unsolvable problem of continuity. Here, the apparatus of probability theory, sets, or information is more suitable. However, the problem is that most of the mathematicians who are attracted to create simulators only know mathematical analysis, so hydrodynamic modeling has gone and continues to go in the wrong direction. There is too much software developed to change anything, and one needs to change it.

The main difference between an adaptive hydrodynamic model and a deterministic one is that the adaptive model is clearly discrete. It does not pretend to work with continuous quantities. A cell in an adaptive model is not a point. This is a part of the space and its main characteristic is the pore volume. Yes, it is assigned one parameter, say normalized hydrodynamic conductivity (the permeability parameter in the model in its pure form is not used at all), but it is understood as the average value with the assumed dispersion, so it does not strictly affect the course of the process. Normalized hydrodynamic conductivity is a dimensionless value in the range from 0 to 1 and it characterizes the probability of transition from one cell to another.

Oil saturation is also the average characteristic of a cell. It should be distinguished from the oil saturation of the rock. Recall that the vertical thickness of the cell can reach 5–10 m. For example, a certain cell can pass through the oil-water contact, but it is still characterized by one average value, although half of the cell is oil-saturated, and the second is saturated with water. Such a cell will still have a non-zero phase permeability, which is determined depending on the oil saturation, even if it is nonlinear. The phase permeability cannot be zero if there is at least some oil in the cell. After all, the flow of oil can always pass through the oil-saturated part of the cell. This probability is never zero. In addition, it is known that the oil-water contact is not a single oil-water interface, but a certain transition zone, the thickness of which is estimated at 5–10 m, where the oil saturation gradually decreases down the section.

In the adaptive model, there is no concept of critical oil saturation, below which oil is stationary. The probability of overflow in this case is very low, but it is never zero. At those low speeds that are observed in the reservoir, the flow is always possible and this is confirmed by the experience of long-term oil fields that are not fully developed for more than a hundred years.

In the adaptive model, the oil saturation can be less than zero. Certainly, this is physically impossible. However, oil saturation in the adaptive model is not a physical quantity. It shows the probability of getting oil from a specific cell, and this probability cannot be zero, even if it is very small and the degree of this smallness just shows a negative value. This situation with negative oil saturation occurs when a well actually takes more oil than is available in its extended area. This is a clear contradiction between the geological and commercial data, but it is not specifically corrected. The problem is that this contradiction arises because the detailed structure of the reservoir remains unknown. Some believe that it may contain high-conductivity channels, about which there is no actual information. The deterministic hydrodynamic model attempts to correct this contradiction by "non-adjacent connections". However, why is it necessary to correct this contradiction? Its presence is more informative than leveling it using any non-physical techniques.

Note here that the weakest point in the deterministic hydrodynamic model is the need to adjust it to the history. However, the hydrodynamic model does not need to be adjusted to the fact at all. Yes, there is a contradiction between the geological model and the field data. However, when creating the geological model, all the information available at that time was used. If contradictions are found, the geological model can be corrected, if the original data allows them to do so. Nevertheless, this should not be done in the process of hydrodynamic modeling, when the permeability is simply adjusted to the field data. This immediately negates all the predictive properties of the hydrodynamic model, since it has already adjusted to a strictly defined forecast by tuning into the fact. Since one cannot feed the same hay to a cow twice, one cannot use the same information twice. One only needs to forecast on a non-fact-based model, then there will be some value in this.

Due to its mathematical apparatus, either liquid rates or depressions are input into the deterministic hydrodynamic model, and everything else must be calculated by iterative solution of a system of equations. This is the clearest manifestation of the defectiveness of the deterministic approach. After all, the rates of liquid and depression are the initial information. Why does not submit all the available information to the model input? This is what is done in the adaptive hydrodynamic model. Oil, water, and gas production for each well for a month is applied to the input. The volume of the month injection upload is also served. In addition, data on reservoir pressure is provided. This is why the

adaptive hydrodynamic model is always tuned to history.

A mechanism resembling cellular automata or lattice Boltzmann is used to calculate the model. It is known that for a hexagonal grid, cellular automata completely reproduce the Navier-Stokes equation [3]. The adaptive model uses exactly the hexagonal grid.

Mechanism of adaptive hydrodynamic model calculation

In a deterministic hydrodynamic model, pressure equations are usually solved first, followed by saturation equations. This is roughly the same as in the adaptive model. First, the pressure fields are calculated for each month of the historical development period and for the forecast period, and then the calculation is based on the saturation or density of current oil reserves.

Let us assume that a certain volume of liquid (V_i) is taken from the well, therefore, from the cell that it opens (or is injected into it). As a result, there is a positive (when pumping) or negative pressure drop (Δp) in this cell, which can be calculated using a well-known formula:

$$\Delta p = \frac{V_i}{V_o \beta}$$

where V_o—pore volume of the cell, β—coefficient of reservoir compressibility.

This is usually a very large unrealistic value. The pore volume of the cell at a grid of 25 m is approximately 500–100 m^3, and a comparable volume of liquid can be extracted from the well in a month. Even if one adjusts for the fact that the selection was made in a month, by default the relaxation coefficient is 0.00033334, the pressure change will still be very large, because the elastic capacity coefficient is a very small value. The relaxation coefficient is selected empirically, while usually not reaching the limit of continuity, at which no change in pressure will occur. It is clear that if there is a flow, then there is a pressure drop. The pressure in the model is normalized to conditionally hydrostatic, so it is close to one before the selection begins. A change calculated using the formula could make the pressure negative. To reduce the pressure jump, a part of the selected liquid should be transferred to the neighboring six cells. It is clear that the liquid came from there (Fig. 6.1).

The redistribution of the liquid in the surrounding six cells is not uniform, but already by weight coefficients, which are estimated by the Darcy's law, written as the following formula:

$$P_i = \eta \frac{f\left(\frac{k_{ph}\Delta p_i}{\mu}\right)}{\sum \left(f\left(\frac{k_{ph}h\Delta p_i}{\mu}\right)\right)}$$

where k_{ph}—phase permeability, h—net pay thickness, ΔP—pressure drop, μ—oil viscosity, η—relaxation coefficient.

In this formula involved the reservoir hydrodynamic conductivity, fluid viscosity and pressure drop. It is noteworthy that the hydrodynamic conductivity and viscosity are both in the numerator and denominator, so they are reduced. In other words, absolute values of filtration parameters are unimportant. We determine the value of the weight function for each of the six cells and divide by the sum of all the cells. Therefore, it becomes known what proportion of the liquid should go to which of the six cells. Therefore, even if the phase permeability is very small, there will still be flows. The proportion of

liquid to be taken from the surrounding cells is calculated from the pressure anomaly, but in one iteration of the calculation, it moves not all, but only part of it. The relaxation coefficient in this case is 0.25 by default, so it takes several iterations to redistribute all the liquid. Studies have shown that the lower this coefficient and the more iterations are performed, the smoother the distribution of the liquid is obtained, but this takes more calculated time.

When some of the liquid was taken from neighboring cells, they in turn have pressure anomalies and therefore they must "take" some of the liquid from their neighbors. Since the movement of the liquid follows the pressure drop, the cells will not take the liquid from the cell that took it from them, and therefore there is no cycling of the algorithm (Fig. 6.2).

The process goes further, covering an increasing number of cells, and gradually degenerates. The more cells in the next "ring", the smaller the volume is taken from them and, accordingly, the smaller the pressure anomaly. The goal of the process is to minimize the following functionality:

$$F = \sum \left((P_{oi} - P_i)\right) \rightarrow \min,$$

where P_{oi}—reservoir pressure in the cell obtained at the initialization stage, P_i—current reservoir pressure in the cell.

This is achieved in six to eight iterations. It is usually not possible to get an exact match with the actual initialization pressure, but this is not necessary, because it was also not perfectly correct, it is enough to approach it within the limits of some error. As a result, the pressure field obtained at the initialization stage is corrected by modeling the process of fluid redistribution. The redistribution itself is a systematic process. At first, only the first ring of cells surrounding the well cell is activated. Then again, starting from the well cell, two rings of cells are activated, then three, and so on until the process degenerates and the last ring of cells cannot take the liquid from anyone, because it itself gave away too little and the pressure drop is

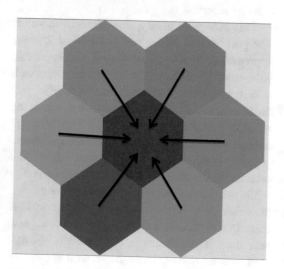

Fig. 6.1 Schematic diagram of liquid redistribution to the well cell

Fig. 6.2 Liquid redistribustion diagram for the next steps

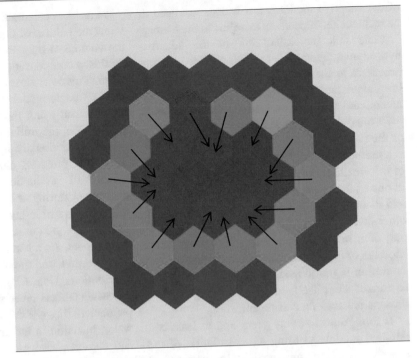

insignificant. By setting the pressure drop boundary for a liquid transition, it is a priori assumed that this liquid is non-Newtonian. Using this systematic structure, the gradual spread of the liquid is simulated. This may seem superfluous, but we know from practice that this is the way to do it. Also, note that despite the fact that the cells are hexagonal, there is still the problem of flow line geometrization. If nothing is done, the anomaly will spread like a hexagon, not a circle. This is less of a problem than with square cells, when one has to deal with diagonals, but one still needs to adjust.

The weight adjustment for transitions between cells is entered, and it changes depending on how many steps are taken from the central cell. The number of steps is calculated during the calculation, and the system always knows it.

The calculation is separated in different phases: oil, water and gas, injection, and it can be calculated in two working agents for injection, for example, water and or water with chemical impurities. For this calculation, one has to keep thirty-three grids in RAM. For example, to calculate for oil, one needs to keep in memory the grid of current distributions for the month and the grid of cumulative production. Four pressure grids are also involved. Two for liquid production and two for injection. Grids differ by one time step of the calculation and if, for example, it is necessary to determine the pressure for the liquid rates, the pressure of the cell where the flow is taken at the grid injection in the previous step, and the pressure of the host cell by grid—for liquid of the current step. The pressures are recalculated after each iteration.

In fact, the calculation of the hydrodynamic model is even more complex, it has a mechanism for displacing oil from the areas of injection wells using the Buckley—Leveret function. There is a disbanding of anomalies during periods of downtime, for example, if a lot of oil was taken from the area of a well, and then it was idle for a long time, then the oil gradually returns to its area. It follows that the calculation of an adaptive hydrodynamic model is not a simplified calculation of a surrogate model. This is a full-fledged calculation that takes a lot of time, which is not much less than the calculation time of the deterministic simulator. However, the advantage of the adaptive model is that it does not need to be recalculated repeatedly to adjust to the history.

Note that the calculation of the adaptive hydrodynamic model is parallelized into as many processes as there are layers in this model. First, the calculation is performed independently by layers, and then the exchange between layers is performed depending on the pressure differences or the material balance of fluids in the cell. This is a moment not only of hydrodynamics, but also of information. For example, it may turn out that in one of the layers too much oil was taken away and its oil saturation became negative, and in the other layer opposite the oil remained a lot, then part of it moves to the layer where there was a shortage of oil. For such movements, negative oil saturation is introduced, so that the system tries to correct them itself. In general, the calculation process is based on self-organization.

It took almost twenty years and at least two hundred models to bring the calculation of the adaptive hydrodynamic model to its current level [4–7]. Each new model can create problems that were not present in previous models, and the algorithm can only be improved on a certain set of models. The main thing is that the calculation of the adaptive hydrodynamic model is stable and always converges. Tests have shown that in the drilled zone, the results of calculating the adaptive model are approximately 75% the same as the results of calculating the deterministic simulator.

Workflow of adaptive hydrodynamic model calculation

The adaptive hydrodynamic model is calculated in three stages. The first two are used for calculating pressure, and the third for calculating saturation. The pressure (P) in the reservoir cell is determined by the following formula:

$$P = \frac{(V_o + V_{zk} - V_{jd})}{(V_o * \beta)}$$

where V_{zk}—injected volume, which came to the area, V_{jd}—fluid volume produced from the area, V_o—total pore volume of the area, β—reservoir compressibility.

In order to calculate the pressure, one needs to know the volume of liquid taken from the cell and the volume of injection that came to the cell. To do this, a compensation model is built at the first stage. This is a simplified single-layer adaptive hydrodynamic model with a grid step of 50 m. It is calculated using the method described above. In this case, the initial pressure is set as normalized to conditionally hydrostatic and close to 1. Then this pressure changes depending on the balance of production and injection volumes distributed during the model calculation. Since the model is single-layer and the grid is enlarged, the calculation is much faster than the full multi-layer model. As a result, we get the fields of liquid, injection, and the actual normalized pressure distributions (Fig. 6.3). During the calculation process, a table is saved for each well with a set of parameters that are calculated in the model: oil, water, injection in the area of the well, pressure, and current phase permeability. There are only nine parameters that are used in the second stage of calculation. The compensation model is calculated for the period of the development history and for the forecast period. How the forecast is calculated for the compensation model will be described in the next chapter.

In the second stage, the pressure fields for each month of the development history are calculated based on the compensation model. This is done based on actual pressure measurements using machine learning methods. The task is to calculate the pressure for each well for each month of its operation. First, dynamic liquid levels in wells are calculated, which are then used to calculate pressures for producing wells and pressures for injection wells.

The number of measurements of dynamic liquid levels in wells is usually much higher than the number of reservoir pressure measurements, and at the same time, the levels reflect the current pressure, so it is advisable to use this information to calculate reservoir pressure fields. However, dynamic liquid levels are not available for every well for every month. It is necessary to fill voids and to calculate the dynamic liquid levels for all months of the operation of the well.

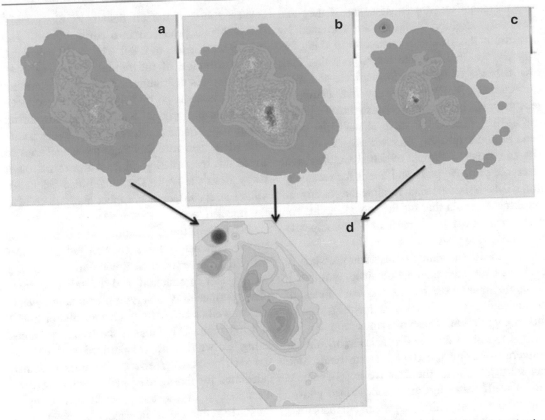

Fig. 6.3 Calculation of the compensation model, **a** distribution of oil production, **b** distribution of water production, **c** distribution of injection, **d** distribution of normilized presuusre

The calculation is performed using a cascade of fuzzy logic matrices. As input data, the results of calculating the compensation model, geological parameters, and directly parameters of the history of the well itself are used. In total, thirty-six parameters are entered in the matrix, nine of which are selected for submission to the calculation. The selection is also made on the principle of minimax of correlation dependencies. The cascade of fuzzy logic matrices has seventy-two layers, thirty-six for averages and thirty-six for variances. It gives a noticeable variation in results. The first point of evaluating the algorithm is that it does not always show the same number and, therefore, responds to changes in input parameters. Because of this calculation, either the

actual measurement of the dynamic liquid level or its calculated value for each well for each month of its operation is obtained, and this is an important input parameter for calculating reservoir pressures.

Reservoir pressure calculations are performed similarly based on actual pressure measurements and based on a compensation model using fuzzy logic matrices. Dynamic liquid levels calculated earlier are added to the parameter matrix. The calculation is performed separately for production and injection wells in order to avoid averaging. For injection wells, dynamic liquid levels are calculated as the average for neighboring production wells. Pressures for injection wells are always a problem, because these measurements are not enough, and if they are, they vary slightly. This can be corrected by measuring the wellhead pressures. In this case, they are recalculated into reservoirs using their own cascade of fuzzy logic matrices, and often this cascade is formed not by one petroleum reservoir, but by a group, because this is a purely mechanical task—how to recalculate the wellhead pressure into the reservoir one. To do this, one can create a sample of actual wellhead and reservoir pressure pairs and use it for recalculation. This is better than recalculating using an empirical formula that fits the same sample. The matrix cascade and the results of the compensation model are also used.

As a result, the values of reservoir pressures in each well for each date are obtained, and thus it is immediately possible to construct fields of normalized reservoir pressures for each date of model calculation. These pressures represent the basis to which the model calculation should converge. Having the base pressure fields, one can start calculating the adaptive hydrodynamic model itself using the method described above. This whole construction may seem cumbersome, but it has elements of the sequential approximation method: first, we build a simplified model, and then a detailed one. In addition, all the actual data on reservoir pressure measurements are used, and thus there is no problem setting up the pressure model. It is always close to actual pressures, just as it is to actual oil and water production.

Adaptive hydrodynamic model always coincides with the fact due to the method of its calculation. The purpose of which is to determine which zones of the oil field most likely came from the oil produced in a particular well and where there is less of it left. The model calculates the distribution of the produced oil (Fig. 6.4), and from this distribution, the density of current reserves is obtained by subtracting the produced ones from the original ones, and the current oil saturation. Therefore, the oil saturation is sometimes less than zero, although the model tries to avoid this by attracting oil from areas far enough away from the well. The model also calculates which reservoir zones are most likely to receive the injected water.

Because of the calculation, we get a table of twelve parameters for each well, which shows the current reserves and oil withdrawals near the well, water production and injection, current phase permeability, reservoir pressure, and water cut, as well as cell activity coefficients for liquid and injection. This table is the basis for calculating further forecasts of production levels and the well work over effect. The model is also calculated for the forecast period, which will be discussed in the next chapter. It is not superfluous to remind that the model is not built for specialists, but for the needs of the forecast system itself. It is possible to visualize at least in order to see the errors of calculation, but the main goal is forecasting. Any inconsistencies and contradictions that occur in it are actually also informative for the forecast.

Here we can quote the words of the great philosopher G. Hegel that contradiction is the criterion of truth, and the absence of contradiction is the criterion of error.

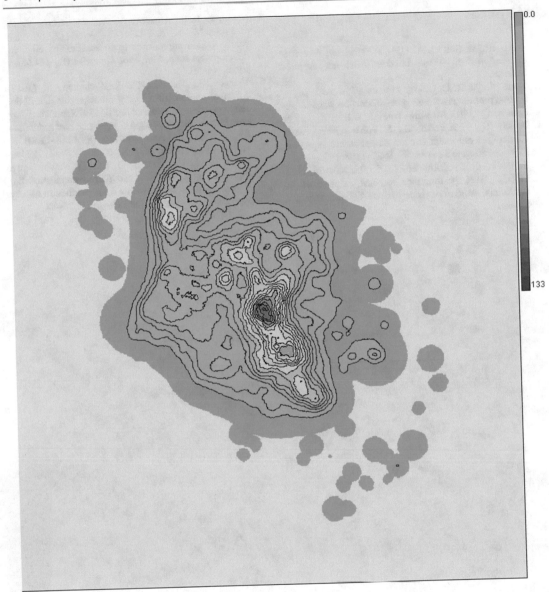

Fig. 6.4 Distribution of cumulative oil production

References

1. Aziz, Kh, & Settari, A. (1986). *Petroleum reservoir simulation* (p. 476). London: Elsevier Applied Science.
2. Kanevskaya, R. D. (2002). *Mathematical modeling of hydrodynamic processes of hydrocarbon field development* (p. 140). Moscow, Izhevsk: ISI.
3. Frisch, U., Hasslacher, B., & Pomeau, Y. (1986). Lattice-gas automata for the navier-stokes equation. *Physical Review Letters, 56,* 1505–1508.
4. Slomatin, G. I., Zakharian, A. Z., & Ashkarin, N. I. (2002). Well performance forecast using artificial neuronets. *Neftyanoe khozyaystvo—Oil Industry, 10,* 84–89.
5. Mandrik, I. E., Guzeev, V. V., Syrtlanov, V. R., Gromov, M. A., & Zakharyan, A. Z. (2009). Neuroinformation approaches to forecasting of oil fields development state. *Neftyanoe khozyaystvo—Oil Industry, 6,* 44–50.
6. Ursegov, S., Taraskin, E., & Zakharian, A. (2017). Adaptive approach for heavy oil reservoir description and simulation. In *Proceedings of SPE Reservoir Characterization and Simulation Conference and Exhibition*, Abu Dhabi, United Arab Emirates (pp. 1238–1250).
7. Taraskin E. N., Gutman, I. S., Rudnev, S. A., Zakharian A. Z., & Ursegov, S. O. (2017). New adaptive approach for the geological and hydrodynamic modeling of petroleum reservoirs with a long production history. *Neftyanoe khozyaystvo—Oil Industry, 6,* 78–83.

Adaptive Forecasting

An adaptive geological and hydrodynamic model is needed not by itself, but as a basis for predicting the future state of the petroleum reservoir development, namely, what oil production will be and what oil well rates will be obtained after the workover measures. The forecast is the least developed area of knowledge. The analysis process is much simpler. There are many works where the methods of system analysis are clearly stated, so each specialist, guided by these works, can independently conduct a competent analysis. It is extremely difficult to find specific literature on forecast issues. As a rule, we are talking about an overly simplified method of forecasting based on regression analysis and the ARIMA algorithm. The complexity of the forecasting method should correspond to the complexity of the system whose behavior it is supposed to predict. It is impossible to predict accurately at all. This follows from Godel's first incompleteness theorem. To make an accurate forecast, it is necessary to go beyond the system under study to a higher level, and this is apparently impossible.

Any forecast made by a computer method is based on information about the previous behavior of the system. However, the future is strictly independent of the past, if only because there are events that did not happen in the past and therefore the total amount of information in the system increases. For example, the early half of the sample for measuring well rates for liquid contains 89,646 lines, 580 values of rates in increments of 1 t/day, and, accordingly, 9.18 bit

of information per measurement with a uniform distribution. The entire sample contains 179,292 rows and 629 values and, accordingly, 9.3 bit of information per measurement with a uniform distribution. In other words, 8.4% of the new values of the liquid rate have appeared, which were not present before, so they cannot be predicted based on previous data. It is noteworthy that the amount of information, taking into account the uneven distribution, even slightly decreased from 6.85 to 6.75 bit per measurement. Regression and all other statistical forecasting methods are ultimately based on distribution, so their forecast is limited. The total amount of information in the system has increased due to new values, and the amount of information for the forecast has decreased.

Back in the early eightieths of the last century, Lorenz [1] showed that non-periodic movement in deterministic dynamical systems (such that the future is uniquely determined by the past) leads to chaos and therefore such systems have a limited forecast horizon. This shows that deterministic methods are not able to predict for the long term. Meanwhile, hydrodynamic simulators calculate oil production forecasts for thirty or more years and their forecast does not become chaotic. This is due to boundary conditions that severely restrict the direction of the forecast and in fact make it meaningless. In other words, a specialist who performed forecasting on a hydrodynamic simulator already had a preliminary forecast obtained using an alternative method, and he or

© The Author(s), under exclusive license to Springer Nature Switzerland AG 2021
S. Ursegov and A. Zakharian, *Adaptive Approach to Petroleum Reservoir Simulation*,
Advances in Oil and Gas Exploration & Production,
https://doi.org/10.1007/978-3-030-67474-8_7

she adjusted his hydrodynamic model to match the results.

This chapter discusses the method of adaptive forecasting of the state of petroleum reservoir development, the effect of well workover measures (WWM) and oil production levels, which is based on almost twenty years of practical experience [2–4].

Any forecast is an inverse problem that does not have a single solution, so in principle it is impossible to successfully predict a single result, such as the flow rate of a single well. However, this does not mean that the forecast does not make sense. The main purpose of this chapter is to show the objective limitations of the forecast based on your experience, as well as to show that in fact, the forecast can be useful even under such restrictions.

The forecast of oil production levels is the final stage of calculating the adaptive hydrodynamic model of the studied reservoir. The calculation includes the formation of four cascades of fuzzy logic matrices of the decline and variation of oil rates, variations of liquid rates, and variations in the injection rates. For deterministic simulators, the same amount of fluid flow or depression is usually applied to the input, assuming that this parameter can be set by purely technological means. It is known from practice that the rates of the liquid and the injection rate vary quite significantly, which is why this variability is taken into account in the adaptive calculation. On the other hand, if only one parameter is sufficient for a deterministic simulator, then for an adaptive model, all three parameters must be input—oil production, water production, and working agent injection.

The forecast of production levels is made iteratively. At the first step, the oil, liquid, and injection rates are set for the forecast period as follows: take the last values by date and stretch forward with a certain average coefficient of oil rate decline, calculated in fact based on the entire sample. Then an adaptive hydrodynamic model for the forecast period is calculated based on the data obtained. Using the simulation results, the predicted oil, liquid, and injection rates are recalculated using cascades of fuzzy logic matrices

in the third step. At the fourth step, the adaptive hydrodynamic model is calculated again. Finally, at the final fifth step, the forecasted oil, liquid, and injection rates are calculated once again. Just five steps, which may seem cumbersome. However, it is known from practice that the forecast oil production is first calculated using liquid displacement curves or oil rate decline curves, and then these results are confirmed using a hydrodynamic simulator. These two fundamental points are combined in an adaptive hydrodynamic model. Moreover, it is not the liquid displacement curves that are used, but the oil rate decline curves, which are the most difficult moment of calculation.

Oil rate decline

When producing one ton of oil, the oil rate in a stationary state, when nothing changes around the well, should naturally decrease, if only because the amount of oil in the reservoir has become less. Various functions, most often hyperbolic, are usually used to predict a decline in oil production. The adaptive hydrodynamic model predicts an oil rate decline based on the numerical solution of the following differential equation:

$$\frac{\partial^2 Q}{\partial t^2} = f\left(Q(t), \frac{\partial Q}{\partial t}, x(t), \ldots z(t) \right)$$

where $Q(t)$ is cumulative oil production, $\partial Q/\partial t$ is oil rate, $x(t),\ldots z(t)$ are parameters of the adaptive hydrodynamic model.

To solve this equation, we select intervals in the history of wells when they did not have any WWM and their flow rate monotonously decreased. The calculation is performed using a cascade of fuzzy logic matrices for nine parameters selected from a matrix of thirty-six parameters using the correlation method. At the same time, such important parameters as cumulative oil production and oil rate are always among the nine selected parameters. This allows to set a cascade of seventy-two matrices, of which half are for averages and the other half are for variances. The calculations performed for several dozen petroleum reservoirs showed that the coefficient of oil rate decline on average is about one ton per thousand tons of oil production and always fluctuates slightly, and often tends to increase (Fig. 7.1).

The decline coefficient is the most important parameter that shows the general trend, but it is known from practice that the graphs of oil, liquid and even injection rates have a pronounced oscillatory character. This shows that the system of a petroleum reservoir is in a dynamic equilibrium, which consists of mutual influences of wells and is expressed in fluctuations of parameters around a certain average value.

Rate variations

It is possible that there is no proven sense in reproducing natural fluctuations of oil, liquid, and injection rates, and this is redundant, but the adaptive hydrodynamic model reproduces them. First of all, in order to complicate the forecast and make it less predictable. If the conclusion of how the forecast was received can be tracked step by step and each step is justified, then this will not be a forecast, but just a calculation that can be performed simply manually. The forecast should be untraceable and unpredictable as a "black box", but with clear rules. All technical products are much simpler than natural ones. Birds are more complex than planes, and fish are more complex than submarines. Analytical solutions that are always too simplistic to reproduce natural phenomena. Therefore, the

forecast system must have elements of self-organization and must be free.

In order to predict variations in rates, the method of time series forecasting is modified somewhat—the future rate depends on the previous value, on the cumulative oil or liquid production, as well as on the parameters of the hydrodynamic model:

$$q(t) = f(q(t-1), Q(t-1), \ldots)$$

where $q(t)$ is current rate, $q(t-1)$ is previous rate, and $Q(t-1)$ is cumulative oil or liquid production.

The forecast is performed using a cascade of fuzzy logic matrices based on nine input parameters. These parameters are selected from a matrix containing thirty-six parameters. A large number of parameters are necessary for the system to have freedom of choice. The choice is made using the minimax method of correlation dependencies. As a result, the values of oil, liquid, and injection rates for each forecast month are obtained. Forecast of oil rates are obtained by summing the results of two calculations: the decline coefficient and variations. In this way, the parameters of oil, water, and injection of the working agent are set for each forecast period, which allows calculating an adaptive hydrodynamic model for this period, and then calculating the same parameters of forecasted oil, liquid, and

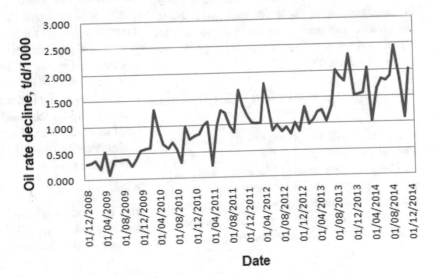

Fig. 7.1 Dynamics of oil rate decline

injection. Since the hydrodynamic model has changed, the parameters have also changed. This could be taken as the final forecasted production levels, but the calculation results need to be strengthened by taking into account well interference.

To do this, a system of equations is compiled in which the rate of oil or liquid of each well at time (t) depends on its previous state at time $(t - 1)$ and from the rates of neighboring production and injection wells for a time (t). These rates are also unknown, so the system of equations is solved iteratively. In addition, the system includes the cumulative production and injection, as well as the current parameters of the hydrodynamic model for the time $(t - 1)$.

$$q^0(t) = f^0\left(q^0(t-1), Q_j^0(t), Q_n^0(t), q^1(t), Q_j^1(t), Q_n^1(t), \ldots q^m(t), Q_j^m(t), Q_n^m(t)\right)$$
$$q^1(t) = f^1\left(q^1(t-1), Q_j^1(t), Q_n^1(t), q^0(t), Q_j^0(t), Q_n^0(t), \ldots q^m(t), Q_j^m(t), Q_n^m(t)\right)$$
$$\ldots$$
$$q^n(t) = f^n\left(q^n(t-1), Q_j^n(t), Q_n^m(t), q^0(t), Q_j^0(t), Q_n^0(t), \ldots q^m(t), Q_j^m(t), Q_n^m(t)\right)$$

There are as many equations in the system as there are operating production and injection wells in a petroleum reservoir. The equations are solved using a fuzzy logic algorithm with nine input parameters. The algorithm is created for each equation.

The equations uses the time series, which are connected to historical rates, obtained after initialization. The result is a slightly different rate, which differs from the initialization. The system begins to correct the predicted initialization rates in the direction of those obtained when calculating with a certain relaxation coefficient. This calculation process is a minimization of the following functionality (F):

$$F = \sum\left((q(t) - q_o(t)) * (q(t) - q_o(t))\right) \rightarrow \min$$

where $q_o(t)$—oil rate before solving the system, $q(t)$—oil rate after solving the system.

The process converges in about 8–32 iterations. In the process, the hydrodynamic model is recalculated periodically every eight iterations. The calculation error is gradually reduced as shown in Fig. 7.2. At the same time, the level of production is changing. Sometimes it increases if the rates of decline were set too high, sometimes it falls on the contrary. As a result, the final forecasted production level is set (Fig. 7.3).

Thus, forecasts of oil and liquid production and injection levels for each production and injection well are justified. This forecast is based on successive approximations. The initial values of rates are set, and the system accepts them based on its history or requires them to be changed. As a result, the rates that the system agrees with are submitted. Summing up the forecasts for all wells, we get a forecast of production and injection levels for the reservoir as a whole (Fig. 7.4).

Usually forecasts are made for a short period of one and a half to two years, but in principle, it is possible to calculate levels for thirty years, just as it is done in deterministic simulators, although one should keep in mind the forecasting horizon and understand that the reliability of forecasts is rapidly decreasing with each month. This is confirmed by the results of the performed testing. Tests were performed many times by cutting off the last year and comparing it with a known fact. Naturally, the forecast accuracy decreases as the forecast period becomes longer. However, as a rule, in the first six months, the correlation coefficients of forecast and actual oil rates are higher than 50% (Fig. 7.5). The correlation for the rates of the liquid is always higher than the rates of the oil, which is also natural.

One can compare these results with a naive forecast if the rate for the next month is set as the rate of the previous month with a certain drop coefficient. Figures 7.6 and 7.7 show that in the first six months of the adaptive model's forecast for oil rates is significantly better than the naive forecast, and by the twelfth month they are closer. For liquid rates, the adaptive model forecast is significantly better than the naive forecast for almost one year.

WWM performance forecasting

This is the most complex type of forecasting by structure, which in principle corresponds to the increased costs that are made during the WWM.

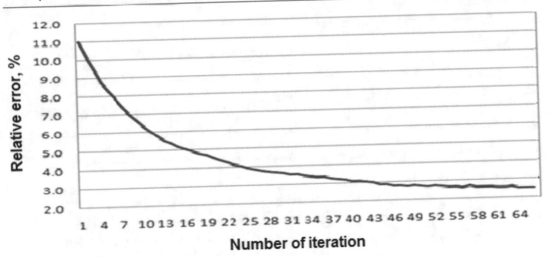

Fig. 7.2 Error minimization

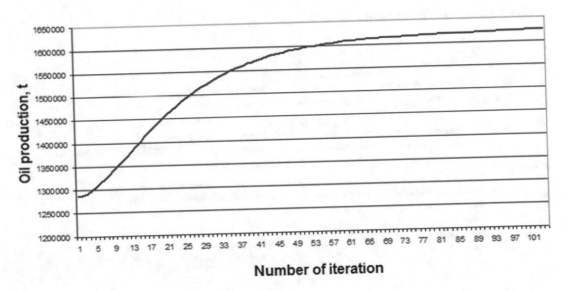

Fig. 7.3 Growth and stabilization of oil production level

The average values of well oil rate, liquid rate and water cut are predicted for three months after the WWM is performed on the well. There is some redundancy here, since these three parameters are interrelated and any one of them can be obtained from the other two. However, each of them is predicted independently, so the forecasts do not converge as they should and this gives additional information about the forecast. It is the oil rate that is predicted, not its growth, because this logically follows from the parameters that can be applied to the calculation input.

Technological parameters of WWM, such as the volume of propane in hydraulic fracturing, are practically not used, because information about these parameters is more difficult to obtain for all WWM. Most importantly, it is not known what the values of these parameters will be during the new WWM. It is not possible to simultaneously predict these parameters and the effect of the WWM itself. Very often, the technological parameters are significantly correlated with the geological and field parameters of the well, because these parameters were taken into account

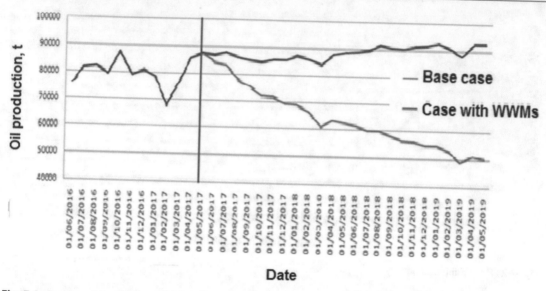

Fig. 7.4 Options of forecasted oil production levels

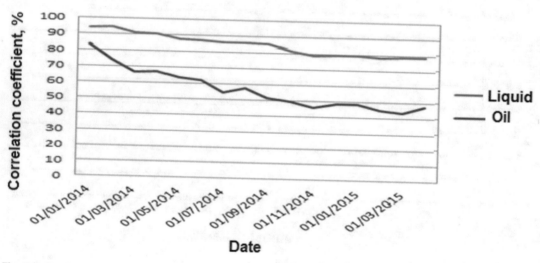

Fig. 7.5 Decline of forecasting accuracy in time

when planning the WWM. For example, the volume of the proppant correlates well with the net pay thickness.

On the other hand, there is no problem in adding technological parameters to the calculation, if the necessary information is available for this purpose. From the experience of using technological parameters of WWM, the effect is known that not always a parameter that is significant for physical reasons, was significant for predicting WWM. The reason is that this parameter may not be effective for statistical processing, either because of its weak variation as a result of a large number of identical values, or because of an inappropriate distribution or a high value of the Lorentz coefficient, i.e. large jumps in the derivative.

Another point is that the true values of oil rates after WWM are not known. We cannot close our eyes to the fact that they can often be

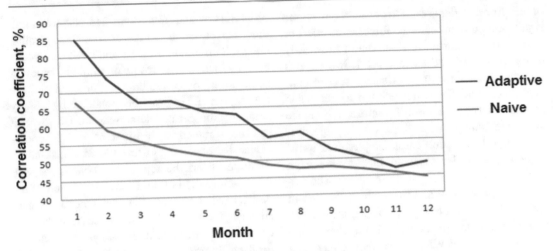

Fig. 7.6 Comparison adaptive and naive forecasting versions for oil rates

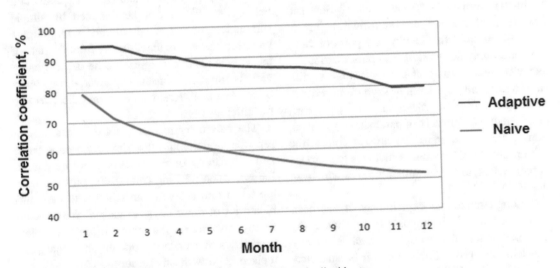

Fig. 7.7 Comparison adaptive and naive forecasting versions for liquid rates

willfully smoothed out. The result is not the true estimated rates and subjective rates. Therefore, it is not very promising to talk about the physical side of the calculation. That is why the probabilistic-statistical approach used to solve this problem is the only possible one, taking into account the uncertainty of the information that is available in fact.

It is predicted that it is possible to obtain an increased oil rate under the geological and field conditions in which the well is located. If these conditions are favorable, there is a high density of remaining reserves and the possibility of recovering them, then any WWM gives a good effect, but if these conditions are bad, then no WWM will help to increase the oil rate of the well.

Artificial neural networks are most often used for this type of prediction. Previously, they were also used in adaptive modeling as a clustered cumulative cascade, but now they have been replaced by cascades of fuzzy logic matrices [5]. In any forecasting method, the first step is to identify the most significant parameters. This is usually done using correlation dependencies. In fact, this is the main point on which the success of the forecast depends. However, there is one significant problem here. To forecast the effect of WWM, one should have a sample of previously made WWM of the same type. However, this sample is never too large. Usually several hundred rows, but this is not enough to get stable correlation dependencies. Therefore, the choice of significant parameters is often subjective, configured for a specific test sample. In adaptive modeling, in order to avoid this, a very large matrix is formed, in which there are about a thousand parameters and from which the system independently selects significant parameters.

Forming a matrix of input parameters is the main task of the WWM forecast, since there is a contradiction between the significance of the parameter from the physical point of view and from the information and statistical points of view. Meaningful from a physical point of view, the option may not work due to poor distribution and vice versa. Different parameters are significant in different petroleum reservoirs. In order to overcome these contradictions, a very large, clearly redundant matrix of parameters is formed, from which the system can always choose the appropriate ones. The matrix includes five groups of parameters, the types and number of which are shown in Table 7.1.

Through numerous experiments, it has been established that it is impossible to choose the ideal parameters that would be suitable for any petroleum reservoir. Each reservoir is characterized by its own significant parameters. Therefore, a large number of combinations are formed with the condition that they correlate as little as possible with each other. This scheme works invariantly. Redundancy and multivariance are

properties of natural systems and adaptive modeling has attempted to create something similar, although on a more modest scale. All parameters are obtained as the result of different types of summations or convolutions. For example, the net pay thickness can be obtained by simply summing the permeable sub-layers. However, it can be summed with different weight coefficients, for example, so that the weight function decreases from the top to the bottom, or vice versa—from the bottom to the top. Which of the options will be more effective is not known in advance for a specific object, so it is best to calculate all three and offer the system to choose for itself. In principle, the system is never wrong. For example, the cumulative oil production before the WWM can be obtained by simple summation, or it can be summed with weights that decrease as one moves away from the date of the WWM. These weights can be defined using various functions: linear, exponential, or hyperbolic. Again, different options may be effective for different objects.

The parameter matrix is formed in a special separate module, and this shows the importance and complexity of the process of its formation. Indeed, when forming parameters, physical laws are taken into account, but along with the combinatorics of these physical values. We are constantly working on the matrix in order to reject parameters that do not work on any object and replace them with other options.

All WWM can be divided into three groups due to their different conditions. The first group is the WWM for the operating wells, i.e. for wells that had a work history of at least six months and were operating at the time of the WWM. The

Table 7.1 Groups of inputs parameters of the main matrix

#	Input parameter group	Number of parameters
1	Geological	273
2	Well production history before WWM	147
3	Surrounding well production history before WWM	345
4	Hydrodynamic model parameters before WWM	231
5	Well working regimes before WWM	14
	Total	1010

second group is WWM for new wells that have been operating for less than six months. Sometimes WWM are done, for example, hydraulic fracturing for new wells. Finally, the third group is WWM for wells that have a history, but they have been out of work for a long period of time (more than six months). This is the least predictable group of WWM, and their unpredictability increases with their downtime.

Each WWM group has its own matrix of parameters. However, for new wells, this matrix does not have parameters for the history of the well operation.

Selecting significant parameters from the general matrix is the second most difficult task in predicting WWM. The choice is made using the minimax method: the average correlation of input parameters with the target is maximized and the correlation between them is minimized. It is clear that if two parameters have a high correlation with the target, then they cannot but correlate with each other.

The forecast is made for a specific type of WWM that has at least 32 actual examples. If the sample is large, more than five hundred rows, then the early WWM are cut off. The calculation is performed using a cascade of fuzzy logic matrices. Moreover, two cascades are formed, one of them for those wells in which a specific type of WWM is carried out, the second - for all types of WWM of the group that includes this well, i.e. for transitioning, new or non-operating wells. The results of the two cascades are summarized.

Twenty-one parameters are fed into each stage. This number is based on tests that have tried from eleven to fifty-one parameters. With twenty-one parameters, the cascade contains two hundred and ten matrices for average values, which gives 228,690 coefficients and the same number for variances. This is a powerful enough system to display any variants that occur in the studied reservoir. However, the calculation is further enhanced by variations, which are obtained as follows: one hundred and twenty-eight significant parameters are selected from the general matrix, and then combinations of twenty-one parameters are randomly selected from them,

and a forecast is calculated for each of them. A variation is obtained, which is shown in Fig. 7.8. It can be used to estimate the probability of obtaining an oil rate not lower than the specified value.

The proposed calculation method is based on extrapolation and on information about previous WWM, so it gives a statistical result. It is impossible in principle to accurately predict the rates of individual wells, but one can choose such wells that will increase the average oil rate after WWM by 10–20%.

It is well known that forecasting methods are most widely developed in the field of economics and business. There are a large number of companies that produce stock robot programs or for solving any business problems, including predicting, for example, whether a new movie will be a success and what portion of viewers will like it. Everyone recognizes that an accurate forecast is impossible in principle, and all achieve statistical success, although a separate event is always predicted. However, in business, the conditions are different, because there are samples of millions of rows, and most importantly, the source data is less contradictory. However, everyone works with a statistical result. Figure 7.9 shows an illustration of the statistical method—it is easy to get the red line (statistical dependence) from the blue points, but it is impossible to get the blue points from the red line. However, the forecasting is considered effective.

Let us look at a sample of fifty-five oil well flow forecasts that were used for WWM and for which the actual values of these flows are known. The correlation coefficient between forecasted and actual rates is 55% (Fig. 7.10). The average actual oil flow rate for the entire sample is 10 t/d. If one now sorts the sample by forecast rates (Fig. 7.11) and divide it into two parts, one can see that the average actual oil rate in the left part of the sample is 6 t/d, and in the right part, where the forecasted rate are higher, it is 13.9 t/d. This is 3.9 t/d or 39% higher than the average oil rate for the entire sample, which was selected with the desire to get the highest possible oil rate. Thus, if one chooses wells with a higher forecasted oil rate

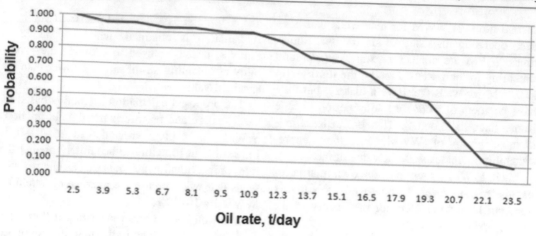

Fig. 7.8 Variation of forecasting

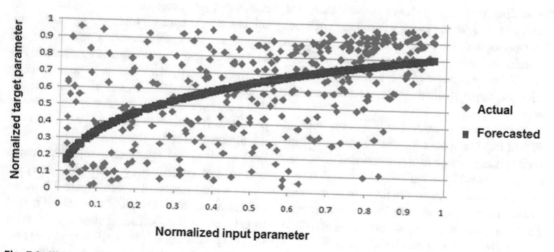

Fig. 7.9 Illustration of statistical forecasting

for WWM, then in general, the actual results of WWM can significantly improve.

Testing is usually carried out in such a way that the history of the last year is cut off, the already known WWM results are predicted and then compared. Despite the evidence, this method is not completely objective. First, if the selection of wells of the last year and the previous time was performed using the same method, then the result will not be predicted at all the effect of WWM, but the effect on already selected wells, so this increases the correlation of the forecasted with the fact. On the other hand, the testing effect depends on a random set of test

wells. Significant parameters are selected based on correlation coefficients with the target parameter for a very limited reference sample of 200–300 rows. In the forecasted sample, the same parameters may have completely different correlations, sometimes of the opposite sign. Therefore, if there is a good match between two sets of correlation coefficients between the reference and test samples, then the test will be successful, otherwise it will not. Figure 7.12 shows a low sample match of 30%, so the test is unlikely to be successful.

Figure 7.13 shows the result of 50 tests of hydraulic fracturing effect forecasts. On average,

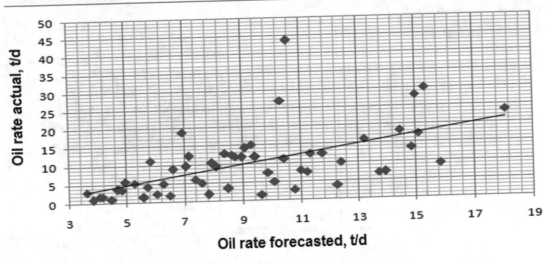

Fig. 7.10 Correlation of forecasted and actual oil rates on test −55%

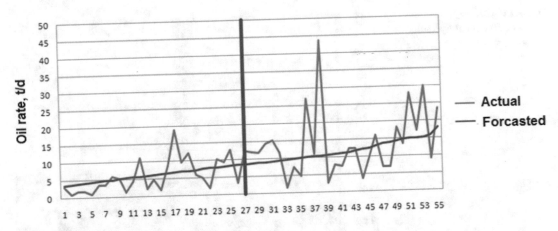

Fig. 7.11 Forecasting effectiveness −30%

the correlation between forecast and actual well flows reaches 56%.

Indeed, retrospective testing on known facts is not entirely reliable. When a fact is known, there is always a risk to adjust the forecasted algorithm to the test sample. Therefore, an attempt was made to obtain the results of more reliable "blind" testing, in which forecasts were made before the WWM was completed. These forecasts were made for 154 wells that were then stimulated. At the same time, hydraulic fracking was also performed for 704 wells for which forecasts were not made (Table 7.2).

The correlation coefficient for 154 wells was only 34%, but the forecast itself was effective.

For 704 for which there was no forecast, the average oil rate was 12.9 t/d. Of the 154 wells, only 45 had a forecasted oil rate higher than 10 t/d, but the actual average oil rate for these 45 wells was 15.9 t/d, i.e. 3 t/d more than for 704 wells. If one moves the section up and take only those wells for which the forecasted oil rate was higher than 15 t/d, the average oil rate will be 18.5 t/d. Thus, testing has shown that if one chooses wells for which the forecasted oil rate is high, one can achieve an increase in the average oil rate after fracturing by 15–25%.

When work on adaptive forecasting of WWM efficiency was started, it seemed that using a large amount of information and powerful

Fig. 7.12 Correlation
between sets of correlation
coefficients of reference and
test samples

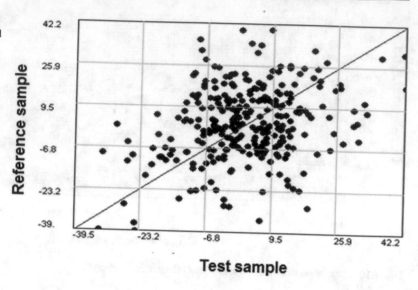

Fig. 7.13 Histogram of
results of 50 tests of hydraulic
fracturing effectiveness
forecasting

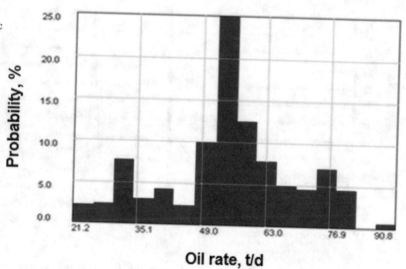

Table 7.2 Results of "blind" testing of the hydraulic fracturing effectivness forecasting

Actual rates	Average for 704 Wells without taking into account forecasting results	Average for 45 Wells where forecasted oil rate higher than 10 t/d	Average for 32 Wells where forecasted oil rate higher than 15 t/d	Average for 22 Wells where forecasted oil rate higher than 20 t/d
Oil rate, t/d	12.9	15.9	18.5	21.1
Additional oil rate, t/d	10.6	11.1	12.6	14.0

artificial neural networks, it was possible to find an algorithm for accurately predicting oil production rates for each well. There was a hope to find such an algorithm. However, after conducting hundreds of tests using various forecasting methods, it became clear that such an algorithm simply does not exist. Indeed, it is known from the theory that the forecast is an inverse problem that does not have a single solution. However, it is one thing to know something theoretically, and another to test it in practice. The main result of the verification was primarily an improvement in the prediction algorithm used. On the other hand, any forecast is not a purely passive matter. If it becomes known and is used, for example, to select wells for WWM, then the correlation coefficients will undoubtedly increase.

References

1. Lorenz, E. N., Nonlinear statistical weather predictions. In *Paper Presented at the 1980 World Meterological Organization Symposium on Probabilistic and Statistical Methods in Weather Forecasting*, Nice.

2. Slomatin, G. I., Zakharian, A. Z., & Ashkarin, N. I. (2002). Well performance forecast using artificial neuronets. *Neftyanoe khozyaystvo—Oil Industry, 10,* 84–89.

3. Taraskin, E. N., Zakharian, A. Z., & Ursegov, S. O. (2017). Adaptive forecasting of well work-over techniques on the example of the permian-carboniferous reservoir of the Usinsk field. *Neftyanoe khozyaystvo—Oil Industry, 7,* 20–25.

4. Taraskin, E. N., Zakharian, A. Z., & Ursegov, S. O. (2018). Adaptive option of steam injection technological efficiency evaluation for carbonate high-viscosity oil reservoir conditions. *Neftyanoe khozyaystvo—Oil Industry, 11,* 102–107.

5. Zakharian, A. Z., & Ursegov, S. O. (2019). From digital to mathematical models: a new look at geological and hydrodynamic modeling of oil and gas fields by means of artificial intelligence. *Neftyanoe khozyaystvo—Oil Industry, 12,* 144–148.

Adaptive Software System Cervart

The method of adaptive geological and hydro-dynamic modeling, which was described in the previous chapters, is implemented in the form of the software system called Cervart. It is an abbreviation of the Italian "cervello artifatto", which means artificial intelligence. The system was officially registered by the Russian Service for Intellectual Property in 2009 (Fig. 8.1), but this does not mean that it is outdated. Over the past years Cervart has been improving on the basis of dozens of the largest Russian oil fields and now there is little left of the program codes of 2009 in this system. At that time, the main tool of the system was a clustered cumulative cascade of artificial neural networks, and at present it is not used at all. Instead, the system developed cascades of fuzzy logic matrices and algorithms for cellular automata.

It is impossible to create a software system like Cervart in one or two years, because it should be tested on several tens or hundreds of objects, since the algorithms of this system are not described anywhere, and they develop together with the system itself. Before the Cervart system, from 2001 to 2009, the Delphor system was developed and tested, which was made on Delphi–7 and had a source database in Oracle. This system was a prototype of Cervart and many ideas were tested on it. However, Cervart is in principle a completely different system, with a higher level of both understanding and implementation. The calculation part of Cervart is created in C++, and visualization and working with the database—in Java. The system mainly runs under the Windows operating system, but it can also be compiled under Unix without any alterations, since it uses only standard C++ functions.

Cervart is generally self-sufficient. It has its own binary database, which contains the model, has ten calculation modules, four modules for loading and preparing source data, as well as a module for exporting simulation results and two visualization modules. This is enough to solve all the functions of the system.

Cervart Structure

To build an adaptive model, Cervart generates a project that is stored in a binary file base on disk. This database is not relational, from which you can make arbitrary selections, but specialized object-oriented and adapted only for the needs of Cervart. During the calculation process, you have to repeatedly access the disk, reading the initial data, writing intermediate and final results. Therefore, this calculation cannot be performed from any external database, such as Oracle. In this regard, like other software packages for modeling oil fields that use local file databases, Cervart also works with its own local database. An adaptive model of a field under study with a thousand wells drilled requires about 2 GB of disk space and the project database contains about 600 files. This shows that the adaptive model has a complex structure.

Figure 8.2 shows the structure view. It is simply located on the computer's disk. Inside the

S. Ursegov and A. Zakharian, *Adaptive Approach to Petroleum Reservoir Simulation*,
Advances in Oil and Gas Exploration & Production,
https://doi.org/10.1007/978-3-030-67474-8_8

Fig. 8.1 Russian state certificate of the Cervart system

main directory, which should have a name in Latin letters and preferably Cervart to avoid confusion, there may be several directories, each for a specific field. Each of these directories can contain several models, each of which is built for one of the layers of the studied field. The internal directory of Cervart is particularly important. It contains all executable (exe) files, projects of auxiliary submodels (901, 902,…) that are used for modeling individual layers of any of the

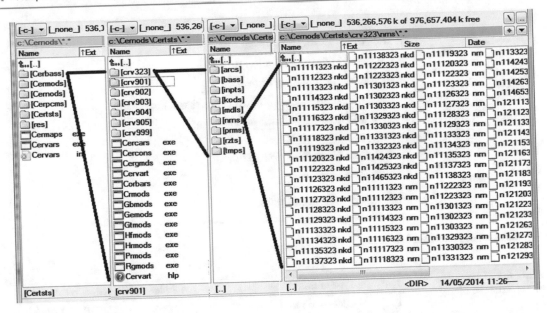

Fig. 8.2 Cervart structure

models. Then there is the crv999 project, a template for creating new projects. However, the main thing is projects crv995, crv997, crv998. These are main projects that combine calculated data from a certain group of oil field models that are somewhat similar. Project crv995—for models of terrigenous deposits, project crv997-for models of carbonate deposits, project crv998-for auxiliary models. It is in these projects that all coefficients of fuzzy logic algorithms or cascades of fuzzy logic matrices collected from all projects are stored. This allows you to "take" ready-made matrices of a similar type of a petroleum reservoir if a certain type of well workover measures, for example, hydraulic fracturing, was not carried out according to some model. In addition, such a project generates cascades of matrices based on aggregated data samples from all reservoirs. In other words, these projects contain the "brain" of Cervart.

There are eight subdirectories within any project, as shown in Fig. 8.3.

All files in the project are binary with a relatively simple structure. First comes the integer code of the file type, then the number of columns, and then the data itself. The file name always starts with a Latin letter, and then two digits show

the file type, the next three digits show the contents, and the last three digits show the project number. The file types are shown in Table 8.1.

This is completely optional for the user. The system itself is guided by the file names. However, the system administrator must know this. In addition, the user may not know all the features of the structure of calculation modules, since their operation is hidden from him.

In total, the system has eleven calculation modules implemented in C++ in an object-oriented style. The user only needs to know the head module—Cervart.exe, which controls the operation of the other eight modules, each of which performs strictly limited functions. The modular system on the one hand facilitates control, and most importantly allows you to painlessly upgrade and expand the functions of each module and the entire system. All calculation modules are console applications managed from the command line. These modules can be compiled for Windows and Linux without any changes (Fig. 8.4).

Head module Cervart.exe manages the entire calculation process-from building a geological model to forecasting production levels and the well workover measure effect.

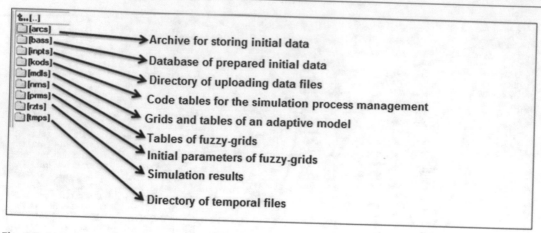

Fig. 8.3 List of project subdirectories

Table 8.1 Types of binary database files

#	Type of files	Code	Name
1	Codes—mixed data type	1,001	k01101999.kds
2	Integer	1,101	n01101999.nkd
3	4-byte real numbers	1,102	b02102999.bss
4	8-byte real numbers	1,103	d11202999.prm
5	8-byte fuzzy-nets	1,106	n11222999.nrn
6	Grid frame	1,111	m12211999.crc
7	Grid	1,112	m12211999.mdl
8	Logging curves	1,115	b04201999.gss
9	List of wells-mixed data type	1,107	b01101999.fnd

It does not count anything by itself, but it runs all the other modules in a given sequence, and also performs parallelization calculations, for example, when creating geological models of individual layers, which are then summed into a multi-layer model.

These calculation sequences are defined inside the module itself. In order to calculate the new model you must first run the command:

Cervart c : /Cervart/Samotlor/crv208 101

for the preparation of the project data.

Then one needs to work a little manually in the module Cermaps.exe. It allows to quickly calculate and visualize the main maps of the studied field: structural, total thickness, and net to gross ratio. One can check and correct strong outliers in stratigraphic markers, including viewing the results of a detailed correlation of neighboring wells. Next, one needs to split the reservoir into layers. This also has to be done half manually. However, after that one needs to run a single command:

Cervart c : /Cervart/Samotlor/crv208 102

in order to calculate the adaptive geological and hydrodynamic models to predict the production levels and the effects of well workover measures.

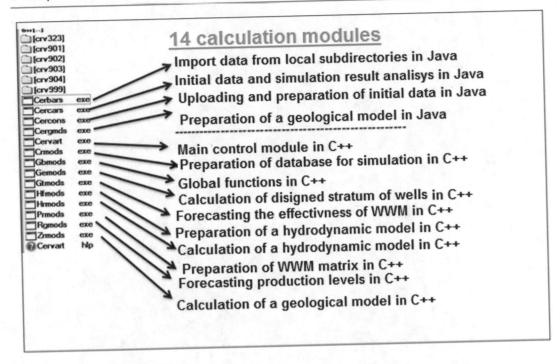

Fig. 8.4 List of Cervart modules

The entire calculation workflow starts with a single command. It is fully automated. A specialist cannot interfere with it because they do not have the technical capability to do so. In general, a specialist is only needed for uploading and verifying data. This cannot be done automatically if there is no reliable centralized database like Oracle. In this case, one can simply export the data. However, most often, the source data has to be downloaded from text files in a special format, which have to be prepared manually. One also needs a specialist in geological modeling, who will build the necessary maps in online mode to check and correct stratigraphic splits that are rarely in perfect condition. Otherwise, all modeling processes are automated (Figs. 8.5, 8.6 and 8.7).

To work with Cervart, one should have at least 8 GB of RAM and an eight-core type i7 processor with a frequency of at least 3.3 GHz. The calculation time for geological and hydrodynamic models, including forecasts of production levels and the WWM effect for a field with 1000 wells and a 40-year development history, is 14–16 h. If the work is related to updating already created models and consists only in adding new data on production and WWM results, it takes 2–4 h for a similar field. A single adaptive model of a petroleum reservoir can take up from 500 MB to 4 GB of disk space, depending on its size and the number of wells drilled. The Java—jre6—jre7 libraries should also be installed on the computer for visualizers and initial data loading modules to work.

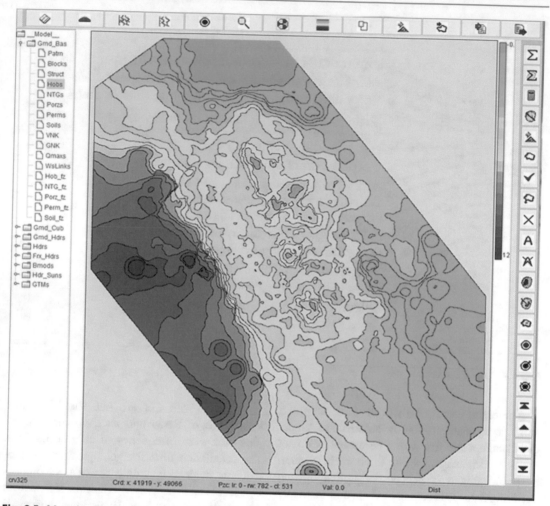

Fig. 8.5 Map visualization and construction module

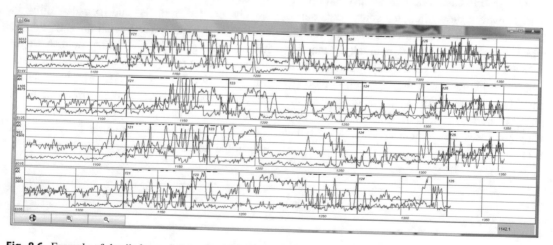

Fig. 8.6 Example of detailed correlation of neighboring wells

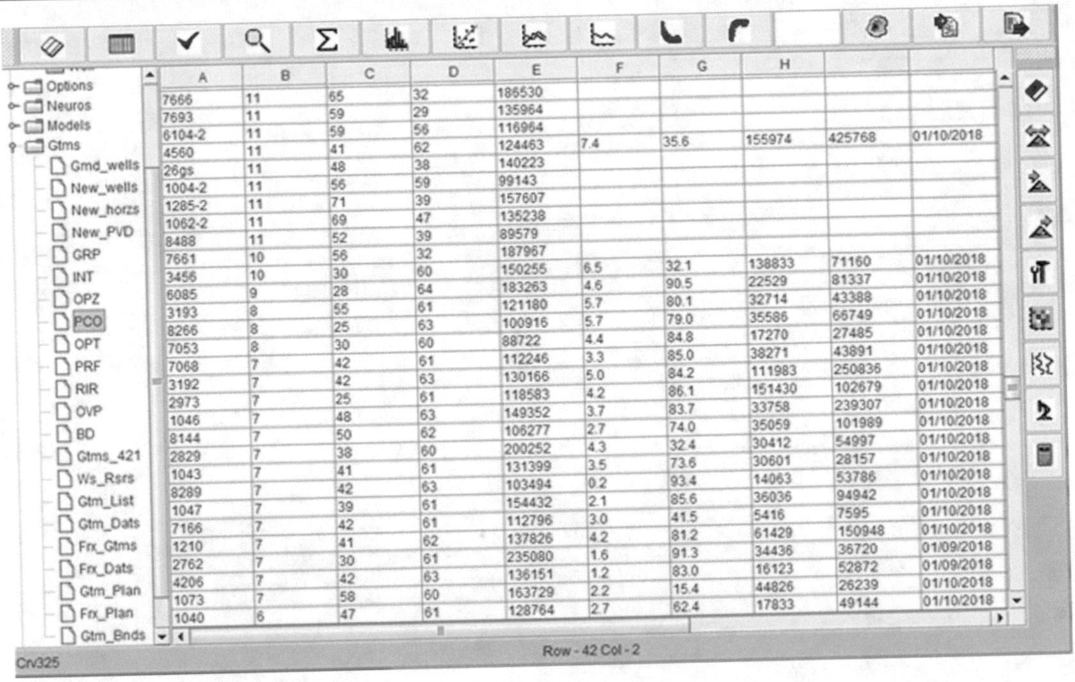

Fig. 8.7 Table visualization module

Individuals who want to obtain a free access to the Cervart system for learning, teaching, training, research, or development purposes can contact Dr. A. Zakharyan by his email: *azarmen51@gmail.com*.

Most of us are hard to convince with reasonable arguments. We do not want to understand, we want to believe, and what is the essence of faith cannot be explained scientifically. Perhaps, it is precisely because there is no reasonable evidence.

Chief executives who make decisions about purchasing deterministic simulators worth several tens or even hundreds of thousands of dollars are unlikely to be familiar with the details of geological and hydrodynamic modeling of petroleum reservoirs. Maybe they want to increase production costs in order to get tax breaks or succumb to advertising. All this mass of books and articles, conferences devoted to geological and hydrodynamic modeling is a kind of advertising. Without it, no one would want to be interested in the advantages of deterministic modeling. Without this, one will not be able to attract well-known experts to his or her side, who will accuse everyone who does not believe in deterministic modeling of heresy.

Who will pay attention to a new book? Thousands of such books are published, and half of them are then recycled into paper in order to print new ones. Any book is always a risk. No one can predict in advance whether the readership will believe in it or not. There are no reliable deterministic or probabilistic methods for this, because the success of a book is a matter of faith in it. Publishers take a risk, expecting to make a profit on the sum of all books published. And they win because they act according to the laws of game theory, perhaps without even thinking about it.

The development of a petroleum reservoir actually also follows the laws of game theory. If one conducts an expensive event, such as hydraulic fracturing, he or she understands that it may be unsuccessful. There can be no reliable method for predicting the result. Kh. Aziz once said that a hydrodynamic model is not needed for forecasting, but in order to better understand the geological structure of a petroleum reservoir [1]. It has been said in this book that a deterministic hydrodynamic model cannot predict because it is usually manually history matched. This adjustment makes forecasting meaningless. It is already done in the process of this adjustment, which inherently ignores the laws of information theory.

The main idea of this book was that the modeling of a petroleum reservoir should first of all rely on these laws, on the amount of initial information that is actually available. Therefore, it is necessary to build adaptive geological and hydrodynamic models that are more in line with game theory.

One needs to focus on fully automated modeling, which will help to avoid unjustified fraud, which is most annoying. At the same time, fraud is often done unintentionally, but with the best intentions. If the simulation system is flexible enough, and deterministic simulators are just that, then it can tune into anything. The algorithm is tailored to a specific situation, but this situation is described by a limited amount of initial data and does not cover all possible outcomes of the petroleum reservoir development system. This is the main contradiction of mathematical computer forecasting, and all modeling is done precisely for the forecast. This forecast cannot but rely on previous information about the development of the reservoir, but this information is contradictory and it is not sufficient for a confident forecast. One can only get a variation of forecasts in a wide range. Deterministic modeling also found a way out, offering many variant calculations. The options are different only by changing the model settings. First, this shows the inherent disadvantage of deterministic modeling. If one can get a significantly different model by simply changing the settings, it means that such modeling is generally meaningless, because the options are obtained not from the variability of the initial data, but from external deterministic conditions or, more precisely, from the contradiction of these conditions to the initial data.

In adaptive modeling, there cannot be many variant models, because there are no settings that can be changed. It is adjusted to the initial data, and forecast variations are obtained precisely from the contradictions between the geological and field data. It seemed that they should be eliminated in the modeling process, but in fact, these contradictions are the most informative, and they allow getting more realistic variations of the forecast.

It is extremely difficult to make serious forecasts. Among the many reasons that negatively affect the accuracy of forecasting, the most important are the large uncertainty and limited amount of available measured data on the properties of the petroleum reservoirs and the technological parameters of individual production and injection wells. That is why, those who need forecasts should always have the means to protect themselves from false assumptions. One of these tools is the proposed adaptive approach of computer forecasting.

The adaptive approach certainly expands the working tools of geological and hydrodynamic modeling, and in some aspects it seems more adequate for computer forecasting than the traditionally used deterministic and stochastic approaches.

From the results presented in this book, it follows that the reliability of adaptive forecasting is primarily due to the fact that this approach uses extrapolation of existing trends in combination with assumptions about the still unrealized consequences of these trends, which may manifest themselves in the short or long term.

The most significant difference between the adaptive approach as a representative of today's popular methods of machine learning and processing big data sets is that for forecasting, the adaptive approach uses multidimensional fuzzy-logical matrices containing about a thousand different parameters, some of which are taken from the adaptive geological and hydrodynamic model, which is necessarily created in automated mode for each object under study and is not much less complex than its deterministic or stochastic counterparts. This allows the adaptive computer system Cervart to be free to choose the most significant parameters and their trends for forecasting.

The main advantage of the adaptive approach seems to be that it has proven to be a working method. This approach has been successfully applied to specific petroleum reservoir engineering problems for almost twenty years.

It is difficult to imagine that all proponents of deterministic modeling can suddenly change their views. Most likely, they will actively protect them. However, everything starts with the first step, and we took that step by creating this book.

Reference

1. Aziz, K. (1989). Ten golden rules for the simulation engineer. JPT, p. 1157.

Index

A
Absolute permeability, 3, 51
Adaptive approach, 84
Adaptive geological model, 38, 49
Adaptive hydrodynamic model, 56, 58
Artificial neural networks, 24, 28, 33, 67

B
Bottom-hole pressures, 11

C
Cellular automata, 34, 35
Cervart, 75, 84
Cybernetics, 23

D
Darcy law, 14, 51
Detailed correlation, 39
Deterministic approach, 1, 8, 19, 20
Deterministic geological model, 37, 49
Deterministic hydrodynamic model, 53
Drainage radius, 3, 15
Dupuit's equation, 3, 14, 23

F
Fluid viscosity, 51, 54
Forecasting, 64, 83
Fuzzy logic function, 31
Fuzzy logic matrices, 33, 63

G
Game theory, 1

H
Hartley's formula, 7
Hydrodynamic conductivity, 15, 43, 47, 52, 54

I
Information theory, 3

L
Lattice Boltzmann method, 34

M
Machine learning, 2, 27
Membership function, 31

N
Net pay thickness, 3, 8, 15, 22, 27, 45, 49, 54
Net-to-gross ratio, 42

O
Oil saturation, 8, 22, 42, 45, 52
Oil viscosity, 54

P
Permeability, 8, 9, 12, 13, 15, 19, 22, 42, 43, 45, 52
Porosity, 8, 42, 45
Probabilistic modeling, 16

R
Reservoir pressure, 3, 9, 10, 12, 54, 56, 57

S
Saturation, 56
Seismic data, 19
Shannon's entropy, 3, 7

U
Upscaling, 37

W
Water cut, 7, 9, 10
Wellbore radius, 3, 15
Well workover measures, 31, 62

Printed in the United States
by Baker & Taylor Publisher Services